Petroleum Discoveries and Government Policy

An Econometric Study of Supply

Petroleum Discoveries and Government Policy

An Econometric Study of Supply

Dennis N. Epple
Carnegie-Mellon University

Ballinger Publishing Company ● Cambridge, Mass.
A Subsidiary of J.B. Lippincott Company

International Standard Book Number: 0–88410–420–6

Library of Congress Catalog Card Number: 75–25626

Printed in the United States of America

Library of Congress Cataloging in Publication Data

Epple, Dennis N
  Petroleum discoveries and government policy.

  Outgrowth of work on the author's Ph.D. dissertation begun at Princeton University in early 1973.
  Bibliography: p.
  1. Petroleum industry and trade—United States—Mathematical models.
2. Gas, Natural—United States—Mathematical models.   I.   Title.
HD9566.E65   338.2'7'2820973   75–25626
ISBN 0–88410–420–6

Dedication

To My Parents

Table of Contents

List of Figures

List of Figures

List of Tables

List of Tables

Preface

This book is an outgrowth of work on my Ph.D. dissertation begun at Princeton University in early 1973. The study investigates an aspect of domestic supply which is crucial to evaluating the long-run prospects for production of petroleum in the United States—the supply of domestic crude oil and natural gas discoveries. The focus of the study is the development and testing of a microeconomic model of producer behavior, a model which takes account of the roles of market incentives and government policies in determining the supply of discoveries. It is hoped that this study will serve as a useful step in the direction of furthering understanding of the functioning of energy markets in the United States.

I would like to thank those persons who provided assistance in the course of work on this study. I owe a particular debt to Stephen M. Goldfeld and Charles H. Berry for advice and encouragement in the development of the work reported here. Richard E. Quandt offered helpful comments and suggestions on several occasions. Michael Visscher read the manuscript and made suggestions which served both to clarify points of the analysis and improve the exposition. Thanks are due Rosaland Tallet who assumed responsibility for typing the manuscript and completed it on schedule. Finally, I thank my wife, Janet, for her unflagging support and encouragement.

Chapter One

Introduction

With the onset of the recent "energy crisis," the importance of petroleum as a source of energy in the United States and all other industrialized economies has become a matter of common knowledge needing little elaboration. Hours of waiting in lines at service stations during the Arab oil embargo have given motorists an intimate familiarity with the connection between crude oil production and gasoline supplies. And all but a small proportion of the American population have seen and heard glowing reports, cast against a backdrop of a wondrously impressive array of imposing but non-polluting machinery and equipment, of the farsighted planning and selfless dedication of oil companies, public utilities, automobile manufacturers, appliance makers, and a host of others to anticipate and satisfy the consumers' energy needs and to provide efficient, economical devices for the productive utilization of that energy.

The current pattern of general public concern about energy supplies is in sharp contrast to the pattern which prevailed previously. In the early years after World War II, the United States was not only self-sufficient in crude oil supplies but was in fact a net exporter. This changed gradually during the fifties as exports declined and imports of crude oil began to rise. Concern, largely by members of industry, was focused not on problems of shortage, but on the price-depressing effects of abundance, and the alleged threat to national security of relying on imports to meet energy needs. The threat to national security (and high prices) was met by the imposition of an oil import quota in 1959.

Considerable excess capacity for crude oil production developed during the early sixties as state agencies found that more and more stringent limitations of production were required to prevent price from falling. Though the domestic price was well above the world price for the late fifties and the sixties, the price was relatively stable with negligible changes for periods as long as five years. Consumer evaluation of the reasonableness of a given commodity price is probably based on the pattern of prices for that commodity experienced in the past rather than on a comparison with foreign prices or marginal production costs. The relative stability of price coupled with the ready availability of petroleum products no doubt explains the relative lack of public concern about petroleum policy.

Though there was little public interest in petroleum policies, the issues were addressed frequently in the economic literature, and there was occasional debate in policy-making bodies, but the status quo was disturbed by policy changes only when events made some type of change inevitable. Most economic analyses have led to conclusions highly critical of U.S. petroleum policies. Federal tax and import policies and state production controls have all been criticized on theoretical grounds for inducing an inefficient allocation of resources. Economic analysts have had no illusions, however, about the impact of these efficiency arguments, as the following comment by Alfred Kahn illustrates: "The economic commentary on the special tax privileges accorded our extractive industries, most importantly oil and gas, has increased sharply in both volume and rigor in recent years, despite the apparent political immovability of the object of all this attention."[1]

With the advent of the "energy crisis" and a growing public suspicion that the petroleum industry bears a good deal of responsibility for events from which it has so obviously benefited, the prospects for displacement of what were once politically immovable objects appear to be quite good. While interest in these traditional policy questions has been rekindled, new policy issues have also arisen. Where judgments concerning previous policies were based primarily on theoretical analysis, these new policy questions require both theoretical and quantitative analysis. Foremost among these is the question of whether the United States should attempt to achieve self-sufficiency in energy supplies.

This is an extremely complex question, and no attempt is made to answer it here. However, a key ingredient required for an answer to this question is a means of estimating future supplies of crude petroleum and natural gas from conventional sources; conventional sources being oil and natural gas discoveries in the "lower 48" states. All domestically produced petroleum is currently obtained from these sources, and they will continue to be the main origin of domestic supplies for the foreseeable future. This study is devoted to estimating supply equations for crude oil and natural gas discoveries from those sources.

The objective is to formulate and estimate a long-run model of crude oil and natural gas discoveries. Properites of particular interest are the long-run price elasticities of supply of crude oil and natural gas discoveries, the elasticity of transformation between crude oil and natural gas discoveries, and the shifting of the supply equations resulting from the exhaustion of discovery opportunities. The equations will yield estimates of the response of discoveries over time to changes in prices or to changes in policy variables (e.g., taxes) which affect price.

An analysis of crude petroleum supply is difficult because of the technical complexity of the supply process. Crude oil and natural gas are joint products in both exploration and production. There are a number of stages in the supply process from preliminary scientific exploration through exploratory and development drilling and production. Oil is an exhaustible resource, and the traditional analysis of competitive supply of reproducible resources must be revised to accommodate exhaustibility. Crude oil production is fraught with externalities, the most significant of which is the problem of multiple ownership of properties overlying a single reservoir. The difficulties in analyzing this technically complex supply process are compounded considerably by state and federal government policies. Government intervention extends from the indirect incentives arising from tax allowances to direct control of imports, production, natural gas prices, and offshore lands. Finally, there are questions about the behavior of firms in the industry, questions linked inextricably to the degree of market concentration and vertical integration in the industry.

OVERVIEW OF THE STUDY

The chapters which follow are devoted to the development of an econometric model of the supply of petroleum discoveries which adequately accounts for the problems identified above. Background information needed for the econometric analysis is provided in Chapter 2. The chapter includes a discussion of the technology of petroleum exploration and production, an evaluation of the available data, an explanation of relevant government policies, and an examination of the salient characteristics of the structure of the petroleum industry. A "back of the envelope" estimate of the price elasticity of crude oil supply based on the analysis of an abstract two-factor production model is developed in Chapter 3.

The first step in the development of the econometric model is the review of prior econometric studies in Chapter 4. A behavioral model of the petroleum exploration firm is developed in Chapter 5, and the equations of the econometric model are derived. Specification of error structures, techniques of estimation, and empirical estimates of the parameters of the model are discussed in Chapter 6. Chapter 6 also contains a summary and interpretation of the results of the study.

The major results of the analysis are based on a non-linear simultaneous equations model. In Chapter 6 various procedures for estimating the parameters of the model are discussed. Alternative estimates of the parameters of the model are discussed in Appendix A, and these results are compared to a Monte Carlo simulation model. Appendix B contains background information concerning the data used in the analysis.

Chapter Two

Technical and Institutional Framework

INTRODUCTION

The activities involved in crude oil production and the nature of technical progress in exploration and production in recent years are described in the first part of this chapter. Sources and reliability of data are then reviewed. The role of federal and state governments in crude oil production and the structure of the crude oil producing industry in the United States are the subject of the latter half of this chapter.

TECHNOLOGY OF PETROLEUM SUPPLY

The major elements of crude oil supply in chronological order are exploration, development, and production. As a background for the econometric analysis, a brief description of what is involved in each of these activities is presented below.

Exploration

Exploratory activities preceding drilling are designed to locate structures which may potentially contain oil or natural gas. Since petroleum is lighter than water, a suitable reservoir must have an impermeable cap to prevent the oil from seeping to the surface under the pressure of subsurface water. Oil or natural gas accumulates in porous rocks or sands beneath such a cap and is held in place from below by porous rock filled with water or by impermeable rock.[1]

Several methods have been developed for locating structures with the characteristics outlined above. Early geological methods involved the study of surface rock structures as a means of inferring the characteristics of the formations below. Later these methods were supplemented by drilling and analysis of core samples.[2] The objective of current geological analysis is to trace the historical development and geographic patterns of various types of formations as an aid in the location of stratigraphic traps where geophysical methods rarely succeed.[3]

Three primary techniques are used in geophysical analysis. Gravitational methods, as the name implies, are used to measure minor gravitational changes to determine the depth and density of subsurface rocks.[4] Magnetic surveys conducted from ground or air measure variations in magnetic mineral content. Seismic refraction and reflection methods involve measurement of shock waves created by surface explosions as they are reflected back to the surface from rock layers below. The time period required for the wave to be returned and the distance from the initial impulse indicate the depth of the reflecting formation and the speed of sound in the intervening strata.[5]

For purposes of this study the details of the various methods are less important than the nature of advancements in exploratory techniques and the type of information which these methods provide. Have technological advancements been largely evolutionary, or have there been major changes in exploration techniques in recent years? If there have been radical changes in the technology of exploration, such changes must be incorporated in the econometric model. Does information gained by geological and geophysical methods deteriorate rapidly in value or is it cumulative in nature? The character of information gained from using these methods will affect the way in which they are treated in the econometric analysis. Fortunately, answers to these questions are readily available.

The National Petroleum Council (NPC) conducted an in-depth study of technological developments affecting the petroleum industry in the period from 1946–65.[6] The study concludes that:

> Developments in exploration technology have been largely evolutionary over the past twenty-year period. While a number of new tools have been developed, the mainstays of exploration are still seismology and subsurface structural mapping, with the use of stratigraphic methods growing rapidly.[7]

In fact, these methods were introduced in the twenties,[8] and their development has been largely evolutionary since that time.

The mappings developed in these studies contribute to a fund of knowledge which may be used over and over again for several years. As a result, the amount of time (crew months) spent in geological and geophysical analysis has declined steadily in the United States since 1956.[9] Geological and geophysical studies are not linked to a particular exploratory well and do not lose their usefulness when a particular well is drilled. These studies thus contribute to the productivity of exploratory drilling in much the same way that technical progress enhances productivity of capital and labor in other industries.

At a fairly early stage in the exploratory process, the oil prospector must obtain a lease to explore on a property. McLean and Haigh provide the following description of the typical lease on privately-owned land:

> The typical oil and gas lease runs for a stated period of time, usually one to ten years, and the landowner is ordinarily paid a cash bonus at the time the lease is negotiated plus an annual rental of a certain amount per acre. The leases frequently contain options with respect to renewals and may or may not contain stipulations with regard to wells to be drilled during the lease period. If production is secured on the property, the annual rentals cease, the landowner secures his royalty interest in the output of the well, and the lease is extended automatically for the life of the well.[10]

Leasing procedures for federal and state lands are described later in this chapter.

If the lease can be obtained and subsequent geological and geophysical analyses are favorable, an exploratory well may be drilled. There is no way of knowing how often a good prospect is rejected, but, of the exploratory wells which are drilled, about one in five is successful.[11] Data collected by the NPC show that the typical oil rig drilled an average of 65,000 feet in 1965. Since the average well depth was 4,380 feet that year, it follows that the average well requires less than a month to drill.[12] Most drilling is done on a contract basis by specialists. Oil well drilling contractors drilled more than 95 percent of all wells in 1958.[13] The average cost per well drilled in 1971 was $94,708, but there are large variations in cost depending on location and depth of drilling. Because offshore wells are deeper and are drilled under more difficult conditions, the average offshore well drilled in 1971 cost $591,196.[14]

As is the case with geological and geophysical methods, technical advances in drilling have been of an evolutionary nature, but have nonetheless been quite rapid. Footage drilled per rig almost doubled in the period from 1950 to 1965, and estimated real cost per foot dropped by 25 percent during that period.[15]

An indication of the outcome of exploratory activities in the United States is provided in figure 2-1 and table 2-1. Figure 2-1 shows estimates made in 1972 of original oil-in-place and recoverable natural gas discovered per year. Total exploratory footage drilled annually is also shown. Table 2-1 gives the distribution of discovery sizes over the last several years as estimated by the American Association of Petroleum Geologists. A rough rule of thumb used by the AAPG for many years was that, to be profitable, a field must yield at least one million barrels of oil reserves or six billion cubic feet of natural gas reserves. Fields in this category are now referred to merely as "significant"; a recognition of the fact that profitability of a field of a given size will change with changes in economic conditions.[16] Nonetheless, this rough rule of thumb is useful in interpreting the figures in table 2-1. The table indicates that the truly enormous fields are occurring with decreasing frequency, but the percentage of fields which are commercially successful appears to have been relatively stable in recent years.

Table 2-1.   Size Distribution of Petroleum Discoveries

Year of Discovery	Number of Fields[a] by Size Group						Total Fields Discovered	Percentage Significant
	A	B	C	D	E	F		
1947	8	12	15	82	164	67	348	33.6
1948	5	5	16	92	276	52	446	26.4
1949	17	11	17	99	215	90	449	32.1
1950	7	14	23	97	327	75	543	26.0
1951	11	5	23	112	402	77	630	24.0
1952	12	9	18	114	426	105	684	22.4
1953	4	9	23	128	372	152	688	23.8
1954	6	4	24	133	412	218	797	21.0
1955	4	8	24	128	436	221	821	20.0
1956	2	6	7	144	425	150	734	21.7
1957	10	6	22	133	444	216	831	20.6
1958	6	8	25	137	372	182	730	24.1
1959	4	5	9	103	390	138	649	18.6
1960	7	5	14	98	418	165	707	17.5
1961	4	4	8	82	402	108	608	16.1
1962	3	5	14	105	425	130	682	18.6
1963	6	2	14	79	465	141	707	14.3
1964	1	3	12	85	466	61	628	16.0
1965	4	3	16	76	372	85	556	17.8
1966	3	2	9	73	422	88	597	14.7
1967	0	2	11	98	430	3	544	20.4
1968	2	1	11	86	339	3	442	22.6

Source: These figures are reported in the *Bulletin of The American Association of Petroleum Geologists,* June issues. The figures are reviewed three and six years after discovery before they are considered final. However, a review of the accuracy of the initial estimates by VanDyke (June 1968) revealed that the initial classifications were generally quite accurate. While the figures for the last two years should be considered tentative, they are probably correct.

Notes:

[a]Size groupings are as follows:
   A: ≥50 million barrels of oil or ≥300 billion cu. ft. of natural gas.
   B: 25–50 million bbls. oil or 150–300 billion cu. ft. gas.
   C: 10–25 million bbls. oil or 60–150 billion cu. ft. gas.
   D: 1–10 million bbls. oil or 6–60 billion cu. ft. gas.
   E: ≤1 million bbls. oil or ≤6 billion cu. ft. gas.
   F denotes abandoned.

Development and Production

If the exploratory well is dry, the prospector goes away sadder but wiser. If oil or gas is discovered, an effort is made to evaluate the size of the reservoir. Additional wells may be drilled to establish its boundaries. If this evaluation is favorable, further development wells are drilled and production begins. The determination of the optimal number of development wells and the optimal production rate for a natural gas reservoir is, in principle, a relatively simple problem. The total amount of natural gas ultimately recovered from a reservoir is, for practical purposes, independent of the production rate or the number of

Figure 2-1. Petroleum Discoveries and Exploratory Drilling (1947-1968)

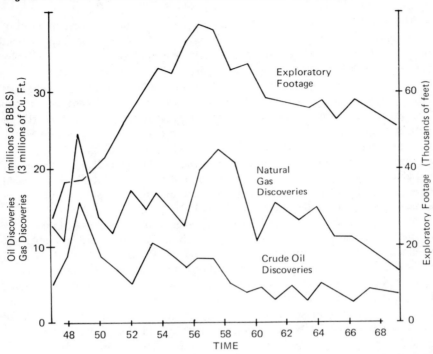

SOURCES: Crude oil and natural gas figures taken from *Reserves of Crude Oil, Natural Gas Liquids, and Natural Gas in the United States and Canada and United States Productive Capacity as of December 31, 1972,* pages 25 and 171. Alaskan discoveries excluded. Footage figures from *Bulletin of The American Association of Petroleum Geologists,* June issues.

development wells used.[17] By adding to the number of wells in the reservoir, the firm can increase the production rate and thereby shift revenues to the present. The optimal number of development wells is reached when the gain in present value from shifting revenues to the present equals the cost of drilling the marginal development well.

In reservoirs containing both oil and natural gas,[18] the recovery process is much more complex because total recovery of oil is dependent on both the rate of production and the number of development wells used. A brief description of the process of displacement of oil from a reservoir will make this dependence clear.[19] At the time of discovery, oil is normally under high pressure from water in the strata surrounding the reservoir. This natural drive pressure forces oil to the surface providing an efficient and economical method for crude oil recovery. Water displaces oil from the pores of rock in the formation and occupies the space thus created. Inevitably, oil in the less porous parts of the formation will be bypassed by this process. The more rapidly production occurs, the greater the amount of oil which will be lost.

The water drive described above is the most efficient natural drive in terms of the proportion of oil recovered, but it may be infeasible if the displacement process is too slow. Alternative natural drive mechanisms may be provided by natural gas in the reservoir. With a gas cap drive, oil is displaced by the expansion of free gas lying above the oil in the formation, while, with the dissolved gas drive, expansion of gas dissolved in the oil serves as the displacement mechanism. The last is the least efficient natural drive, and it is used only as a last resort. For both the water drive and the gas cap drive, ultimate recovery is dependent on the rate of production. However, the loss in ultimate recovery is less if the production rate is increased by increasing the number of wells in the reservoir rather than by increasing the rate of production per well. McKie and McDonald have provided a geometric analysis of the determination of optimal well density and production rate under these conditions.[20]

Most oil produced in the United States comes from fields operated under one or more of the natural drives described above. Methods are available for enhancing production from natural drives and for attaining additional oil when natural drives are exhausted. These methods, termed pressure maintenance or secondary recovery, involve injecting natural gas or water into the producing formation to sustain or replace natural drive pressures. The proportion of total U.S. production from secondary recovery projects was about 15 percent in 1950, 30 percent in 1960, and 28 percent in 1970.[21] Recovery may be further increased by injecting fluids miscible with oil as a displacing agent. For high viscosity oil, thermal methods may be used. With currently used methods, heat is obtained either by burning oil and injected air in the reservoir, or by injecting steam.[22] These methods are relatively expensive and account for a modest proportion of current U.S. production. Kuller and Cummins have generalized the earlier analysis of McKie and McDonald to show how optimal investment and production rates may be determined when natural drive can be supplemented by other recovery methods at any time during the life of the reservoir.[23]

Technological advances in development and production have primarily been in the area of improved understanding of reservoir mechanics and refinements in recovery methods. Cumulative recovery efficiency, defined as the proportion of oil in all known fields which is recovered, provides a useful index of improvements in production methods. The NPC estimates that cumulative recovery efficiency was 17 percent in 1920, 26 percent in 1945, and 36 percent in 1965.[24] While this may appear to be a modest rate of increase, it must be recognized that this gain applies to all previously discovered fields. A 1 percent gain in recovery efficiency results in the production of about 3.5 billion barrels of oil, more than twice the amount of recoverable oil discovered per year in the "lower 48" United States in the last decade.

It appears that improved recovery methods have been applied with equal success to old and new reservoirs. Using the more conservative ultimate recovery figures of the American Petroleum Institute, the cumulative recovery efficiency in 1969 of fields discovered prior to 1935 was estimated to be 30.3 percent while the cumulative recovery efficiency for all fields discovered prior to 1969 was estimated to be 30.2 percent.[25] The NPC estimated that ultimate recovery rates of 50 to 60 percent might be attainable though they were vague about what conditions were assumed in that estimate.[26] It seems fair to say that current technology makes it feasible to achieve those rates, but no careful study has been done to indicate what recovery rates could be expected at various price levels.

It should be clear from this discussion of production that exploitation of a reservoir takes several years. The rate of production is normally highest early in the life of a reservoir, and it declines gradually over time. Evidence is, in fact, available about reservoir decline rates encountered in practice. This evidence will be reviewed here as a background for later analysis. Published economic studies of the petroleum industry have utilized the assumption that reservoir production declines at an exponential rate,[27] and this study will do so as well.

The primary reason that reservoir production rates decline over time is the reduction in natural drive pressure as oil is removed. Using relatively simple assumptions about pressure behavior, Bradley was able to derive a production equation which is exponential.[28] In a somewhat dated review of evidence on decline curves for U.S. reservoirs, Arps asserts that the hyperbolic decline curve is actually a more accurate characterization of decline curves for most reservoirs than the exponential.[29] The superiority of one over the other is not particularly pronounced, however, and Arps observes that many companies use an exponential approximation even when the decline curve is hyperbolic.[30] The differences are most notable for the late stages of production of a reservoir.[a] Since use of the decline curve in this study will be for the purpose of estimating the present value of newly discovered reservoirs, minor differences in production trends late in the life of a reservoir will be inconsequential due to the effects of discounting. Bradley tested both functional forms in his analysis and found that the exponential form adequately approximated production characteristics of reservoirs in foreign fields and that the hyperbolic did not yield an improvement over the exponential form.[31] Use of the exponential decline in economic studies is motivated in part by expediency, but the above evidence suggests that the approximation is likely to be reasonably good. Currently estimated decline rates are on the order of 10 to 12 percent per year.[32] Thus it takes six or seven years after development wells are drilled to produce the first half of the recoverable oil in a reservoir.

TERMINOLOGY AND DATA

The most comprehensive data on drilling of oil and gas wells is collected by the American Association of Petroleum Geologists (AAPG). A well drilled "to

[a]The exponential curve is $q(t) = q_0 e^{-dt}$. The hyperbolic is $q(t) = q_0 (1 + at)^{-b}$.

exploit a hydrocarbon accumulation discovered by previous drilling" is termed a development well.[33] Exploratory wells are separated into several categories on the basis of the relative riskiness of the undertaking. Wells drilled in the vicinity of a known reservoir are distinguished by depth and relative distance from the known reservoir. Wells drilled to a lesser or greater depth than the known reservoir are termed shallower-pool and deeper-pool tests respectively. Wells drilled relatively close to the reservoir are outposts (or extension tests), while wells drilled beyond the limits of the reservoir are new-pool wildcats. In areas not previously productive, exploratory wells are termed new field wildcats.[34]

Estimates of discoveries, reserves, and production of crude oil and natural gas are reported annually.[35] A period of four or five years after discovery is required before development drilling and production experience yield sufficient information about the discovery to provide a reasonably accurate estimate of the original oil-in-place. Proved reserves are an estimate of the amount of oil ". . . which geological and engineering data demonstrate with reasonable certainty to be recoverable in future years from known reservoirs under existing economic and operating conditions."[36] Ideally, proved reserves are a cumulative total of recoverable oil (or gas) discovered less total cumulative production to date. Published estimates are deliberately conservative as the term "reasonable certainty" indicates. Thus, for the first few years after discovery, the estimate of reserves in a reservoir is based on the proportion of the reservoir which is developed and producing—not on anticipated ultimate recovery from the reservoir. Total reserves are reported each year but are not broken down by year of discovery. Estimated ultimate recovery for fields discovered each year is also reported. This is defined as cumulative production from those fields plus remaining proved reserves in those fields. As indicated previously, ultimate recovery to date has been about 30 percent of original oil-in-place. Production data are also reported on an annual basis but not attributed to the year of discovery of the field from which the oil is taken.

Almost all natural gas in a reservoir is eventually recovered so the distinction between gas-in-place and ultimate recovery is unimportant, and only estimated ultimate recovery is reported.[37] Natural gas volumes are measured at a standard temperature and pressure, and adjustments are made for removal of liquids which may be dissolved in the gas. The amount of liquids recovered from the gas is also reported.

The U.S. Bureau of Mines reports crude oil and natural gas production and prices at the wellhead.[38] Production data reported by the Bureau of Mines are obtained independently of data reported by the American Petroleum Institute. The differences between the two estimates are normally small—less than one-tenth of one percent in 1972.[39] American Petroleum Institute data are used in this study. Indices of industry drilling costs have been compiled by the Independent Petroleum Association of America (IPAA) since 1947.[40] Costs of drilling and equipping wells classified by depth and type (oil, gas, or dry) have been reported by the Joint Association Survey for 1953, 1955, 1956, and annually since 1959.[41] The IPAA index is used in this study since it extends over a longer time period. This index is designed to reflect changes in aggregate drilling costs. It is a weighting of the costs of over twenty inputs used in drilling, and it is adjusted to reflect changes in the depth of wells.

Reliability of the Data

With the exception of production and prices, all of the data identified above are prepared by industry sources. This pattern of governmental reliance on the petroleum industry for data (and frequently for analysis as well) is not new, nor is it limited only to the producing stage of the industry. For example, the study of technology in the petroleum industry cited earlier was prepared by the National Petroleum Council at the request of the Department of the Interior.[42]

Does it really matter who prepares the data? One might argue that, as a matter of principle, the government ought not be dependent on industry for data or analysis, but an examination of questions of principle would not serve any purpose for the present study. From a practical standpoint, it matters if the source of the data affects what is reported or the accuracy of the reports.

Regarding the former issue, it is undoubtedly true that the types of information and reporting procedures used by industry organizations such as the National Petroleum Council, The American Petroleum Institute, etc., are designed to minimize disclosure of information about individual firms and to place the industry in a favorable light. Data on discoveries, reserves, and production are reported on a geographic basis, but no information for firms or even selected groups of firms are reported. Similarly, wells are classified by depth, by degree of risk, by outcome (oil, gas, or dry), and by cost; but the role of firms of different sizes or different degrees of vertical integration is not reported. One could hardly expect that data would be reported on a systematic basis for each firm, but the degree of aggregation in the reporting of industry data goes well beyond that required to prevent disclosure of information about individual firms.

Concerning the issue of accuracy, there is no certain means of determining the reliability of the data which are reported. There can be no doubt, however, that a good deal of effort has gone into developing the concepts and procedures used in preparing data on reserves, discoveries, and drilling. One cannot read an examination of these issues, such as that by Lovejoy and Homan,[43] without being convinced that a sincere effort has been made to provide reliable information.

There do not appear to be any incentives for the individual firms to provide false information concerning discoveries, reserves, production, or drilling because they are granted complete anonymity in the publication of industry statistics. Firms may be careless in providing information, but that is a problem which characterizes all data gathering activities that rely on voluntary compliance.

Since the advent of the "energy crisis" of 1973-74, accusations that the industry has provided misleading information have proliferated in the popular press. For the most part, these outcries have concerned reporting of inventories of fuel oil and gasoline during the "shortage." The circumstances under which these inventory data were reported differ in significant respects from those under which the data used in this study are reported. Inventory data were being collected by the government for the express purpose of determining allocations to retail distributors at a time when prices were controlled administratively. Shortages occur because price is not sufficiently high to clear the market, and that point had certainly not escaped the attention of firms holding inventories of refined products. With the expectation that prices were going to be allowed to rise, firms with storage capacity would naturally have preferred to hold their inventories until the price rise had been realized rather than report them to the government and have them allocated for sale in the intervening period. Under those circumstances, the incentives for reporting false information were enormous, and it would require a considerable act of faith to accept that industry data at face value. There will be no occasion for using such inventory data in the analysis in the following chapters.

It is not the purpose of this study to determine what petroleum industry data should be collected or by whom. A cursory glance suggests that the industry, acting as its own census taker, has reported information which meets the needs of its members or the requests of government in a way which carefully conceals the role of firms in industry activities. This is a hindrance to economic analysis of behavior of firms in the industry, and it is probably intended to be. However, it is doubtful that anyone doing empirical analysis has ever gotten data in the form and detail he would like, and considerable headway can be made with the data that are available.

Data Used in the Econometric Analysis

This study is concerned with estimating supply equations for crude oil and natural gas discoveries. In the course of the analysis, data on original oil-in-place, ultimately recoverable oil and natural gas, and production of crude oil and natural gas will be used. In addition, data on the number and depth of exploratory wells drilled annually, and the IPAA index of drilling costs will be used. Price data prepared by the Bureau of Mines will also be used. The available evidence suggests that this information has been collected systematically with care being taken to develop meaningful measures of industry activity. Since there is no evidence that this information has been falsely reported, and there do not appear to be incentives for firms to supply misleading information, this study proceeds on the assumption that the data are reliable and accurate.

GOVERNMENTAL POLICIES

Government intervention at both the state and national level has been an important factor in the domestic supply of petroleum for virtually the entire history of the industry. A description of major policies follows.

State Government Regulations[44]

It would be difficult to find a discussion from an economic viewpoint of state regulation of the oil industry which does not begin with reference to the rule of capture. The rule of capture provides that oil brought to the surface on a piece of property belongs to the property owner.[b] Since oil moves beneath the surface without respect for property boundaries, each property owner has an incentive, under the rule of capture, to drill wells as rapidly as possible and to produce oil as quickly as possible.

In principle, the solution to the problem of underground migration is obvious; the pool should be operated as a unit. Homan and Lovejoy have commented, "If development or production activity were not permitted to proceed until an agreement on unit operation had been reached, no doubt a great capacity for voluntary cooperation would be revealed."[46] In practice this has not been done. Instead, the major producing states have stepped in to regulate the industry. Most state regulations were developed during the early thirties at a time when major new oilfields were being discovered in Oklahoma and Texas, and crude oil prices were falling precipitously. It is problematic whether the impetus for state regulation derived primarily from concern about economic inefficiency and waste or from concern about large financial losses being sustained by the oil producers.

In any case, state regulatory agencies assumed a major role in managing the affairs of the oil industry. Well spacing, allowable production rates, gas-oil production ratios, surface storage methods, waste disposal, pressure maintenance, reasonable prices, employment, and equity among producers and property owners are among the major issues with which the regulatory agencies are concerned. The most controversial activity of the regulatory commissions is the prorationing of output to market demand.

The prorationing procedures used by the major producing states differ in detail, but their broad outlines are essentially the same. The procedures used by the Texas Railroad Commission will serve as an illustration. A schedule of allowable production has been established by the commission giving the maximum number of barrels per day which can be produced from a given well. This "yardstick" is dependent on the depth of the well and the spacing of wells

[b]The rule was established in court decisions in various states in the 1880s before the importance of underground migration of petroleum had been established.[45]

in the field. The schedule of allowables appears to have been established primarily for administrative convenience since principles of conservation and economic efficiency have not been clearly formulated by the regulatory agencies. The yardstick is intended to specify the maximum amount which the well would be allowed to produce in the absence of market demand restrictions.

The output to be prorated among the producers for a given month is determined as follows. The Bureau of Mines provides an estimate of demand at the current price. The state supplements this with its own estimates. Purchasers of oil in the state are required to submit sworn nominations indicating their intended purchases during the coming month. The state estimates the amount of oil to be expected from other sources—primarily imports and changes in inventories in surface storage. From its estimate of demand the state subtracts exempt production[c] and oil from other sources. The remaining output is allocated among the non-exempt wells. The allocation to these restricted wells is normally specified as the number of production days allowed during the coming month. The operator of a particular well then knows that the coming month's output from that well cannot exceed its yardstick times the number of production days allowed.

The regulatory agencies generally claim to ignore price in establishing market demand. This claim has been viewed with considerable skepticism by economists. DeChazeau and Kahn made a thorough review of the events leading to price changes in the fifties. They concluded that price was effectively set by the regulatory agencies through their decisions to ratify increases in posted prices by imposing the necessary restraint on production. They document instances in which increases in the posted price were ratified by state production controls as well as instances in which price increases were rescinded when the requisite reduction in output was not mandated by the state agencies.[47]

Interstate Allocation of Production. No currently available study provides a completely adequate explanation of the procedures by which production is allocated among the states. Three alternative explanations of the allocation process have been offered at one time or another. The first is that allocations are determined at the semiannual meetings of the Interstate Oil Compact Commission (IOCC).[d] Limitation of production for the purpose of fixing price is explicitly rejected in Article V of the agreement creating the IOCC.[48] However, Lovejoy and Homan have suggested that the semiannual meetings of

---

[c]A brief characterization of exempt production is difficult. Discovery wells are exempted for a period of time to encourage exploration. Most other exempt wells are low volume wells on artificial lift with relatively high recovery cost per barrel. Exempt wells have produced a substantial proportion of the total in the past.

[d]Thirty oil and gas producing states belong to the IOCC, but the largest five are responsible for more than three-fourths of U.S. production. See table 2–3.

the IOCC provide a forum both for discussing regulations limiting output and for maintaining a general awareness among the states of what the others are doing.[49]

A second explanation is that allocations are determined by the monthly forecasts of state demand distributed by the Bureau of Mines. A 1957 report by the attorney general contains a study of the role of the Bureau in the interstate allocation process.[50] The report states that: "It is generally agreed that state prorationing would have been difficult, if not impossible, to administer without a comprehensive body of statistical data comparable to that furnished by the Bureau."[51] After a review of the accuracy of the monthly Bureau of Mines estimates relative to actual consumption over a period of several years, the study concludes that these forecasts are not sufficiently accurate to serve as a basis for state production restrictions. They do conclude, however, that the weekly reports of changes in stocks are a key item needed by the states in preparing their own forecasts.[52]

Finally, it has been suggested that state production controls bear a striking similarity to patterns observed in oligopolistic industries when price is held substantially above marginal cost by dominant firms. Maintenance of the artificially high price requires that the dominant firms curtail their own production as small firms expand and new firms enter under the price umbrella created by the large firms. Adelman makes this case forcefully: "Texas could permit greater production and get more of the market, but that would wreck prices. This is the classic dilemma of the largest producer in a cartel, who holds the umbrella for those nibbling away at its market share."[53]

While the mechanics of interstate allocation must remain an open question for the time being, table 2-2 provides incontestible proof that production was severely restricted for a period of several years in the face of stable or rising prices. The excess capacity figures reported in table 2-2 are only rough estimates since not all reservoirs would necessarily have been operated at capacity even in the absence of state regulations. Still, these figures give a reasonably accurate portrayal of the pattern of excess capacity for the period shown.

A comparison of the pattern of excess capacity across states to the distribution of production across states (table 2-3) provides considerable support for Adelman's position that the large states have borne the major burden of restricting output to maintain price. The states with the largest shares of production have the highest percentages of excess capacity. It appears that producers in smaller states benefited from the higher prices, but that those states contributed very little in the way of restricting production to maintain that price.

Current Status of Prorationing. Excess capacity has characterized the production of crude oil during the entire period for which data for this study are taken (1947-68). It is for this reason that the mechanics of prorationing have

Table 2-2.   Crude Oil Price Per Barrel and Estimated Excess Capacity

Year	Percentage Excess Capacity	Average Price per Barrel	Posted Price per Barrel
1947	NA	1.93	1.62
1948	NA	2.60	2.57
1949	NA	2.54	2.57
1950	NA	2.51	2.57
1951	NA	2.53	2.57
1952	NA	2.53	2.57
1953	NA	2.68	2.57
1954	27.3	2.78	2.82
1955	27.1	2.77	2.82
1956	27.4	2.79	2.82
1957	31.9	3.09	2.82
1958	45.8	3.01	3.07
1959	41.3	2.90	3.00
1960	43.3	2.88	2.97
1961	44.3	2.89	2.97
1962	44.8	2.90	2.97
1963	42.6	2.89	2.97
1964	42.5	2.88	2.97
1965	42.6	2.86	2.97
1966	37.1	2.88	2.97
1967	32.8	2.92	2.97
1968	30.5	2.94	3.07
1969	28.3	3.09	3.07
1970	21.7	3.18	NA
1971	21.6	3.38	NA
1972	NA	3.40	NA

Sources: The Independent Petroleum Association of America made annual estimates of the capacity for crude oil production with existing facilities and equipment at the beginning of each year for the period from 1954 through 1972. Figures are reported in annual issues of the IPAA publication *The Oil Producing Industry in Your State.* Excess capacity figures in the table are the amount by which productive capacity exceeded actual production during the preceding year. For example, productive capacity on January 1, 1972 exceeded average daily production in 1971 by 21.6 percent. Remaining figures in the table are average price taken from *The Oil Producing Industry in Your State,* 1973, p. 93, and posted price taken from *Petroleum Facts and Figures* (Washington: American Petroleum Institute, 1971), p. 449.

been reviewed in this chapter. In recent years, the importance of prorationing has slowly declined, and, at present, no output restrictions are imposed beyond the maximum allowable discussed earlier. When excess demand prevails, prorationing is unnecessary, and upward price adjustments are made without state intervention. It now appears unlikely that prorationing will be a significant force in the future.[54] Import controls have been changed from a quota to a tariff thereby linking the U.S. price to the world price. Without the quota, state controls will be ineffectual since no state has a large enough share of world production to affect price. It is conceivable that a rapid increase in prices in the short run might be followed by a substantial supply response as longer run

Table 2-3.   Excess Capacity and Percentage of U.S. Production by State

State	Excess Capacity 1965	Excess Capacity 1971	Percentage of U.S. Production 1965
Texas	69.0	24.8	34.7
Louisiana	74.7	34.6	19.4
California	13.8	15.1	11.7
Oklahoma	13.9	5.5	7.6
New Mexico	13.0	12.8	4.3
Kansas	10.9	6.2	3.8
Total	42.6	21.7	81.6

Source: The Independent Petroleum Association of America made annual estimates of the capacity for crude oil production with existing facilities and equipment at the beginning of each year for the period from 1954 through 1972. Figures are reported in annual issues of the IPAA publication *The Oil Producing Industry in Your State.* Excess capacity figures in the table are the amount by which productive capacity exceeded actual production during the preceding year.

adjustments are made. If this were accompanied by a collapse of the Middle Eastern cartel and a rapid reduction in world oil prices, the state and federal governments might once again decide to support the U.S. price with import and production restrictions. Needless to say, these are questions beyond the scope of this study.

Leasing of State Lands.[55] The states have been granted control of offshore lands up to three miles from shore with the exceptions of Florida and Texas which control nine nautical miles from their respective Gulf Coasts. Many states have claimed land beyond the three-mile limit and some of these claims are being contested in federal courts. Though Florida and Texas were successful, the three-mile limit established by federal law has been upheld in most cases.

The states have not adopted uniform regulations governing exploration and leasing of offshore lands. Most state regulations have similar provisions, however. For preliminary exploratory activities, most states require application for a permit and payment of a modest filing fee. Leasing of lands is normally by competitive bidding either at state initiative or at the request of interested parties. State laws typically require a royalty of from one-eighth to one-sixth. Bids of cash bonuses and/or annual rentals are then the basis for determining who will obtain lease rights. All states reserve the option of rejecting bids for part or all of the properties for which bids are requested. The initial term of leases is from three to five years in most states unless a discovery is made in which case the lease is automatically extended.

Federal Government Policies

Major federal government policies affecting the petroleum industry are import policy, tax policies, federal land leasing policies, and regulation of natural gas prices.

Import Policy. The federal government initiated a voluntary program to limit imports in 1955. The voluntary program was ineffective from the outset, and eventually collapsed as companies evaded or ignored import allocations. The divergence between domestic and foreign prices, estimated to be on the order of 20 percent in 1957,[56] was a stronger influence than government appeals for voluntary cooperation.

In 1959 a mandatory program was established limiting imports to 9 percent of estimated domestic demand. This was changed to 12.2 percent of domestic production in 1962.[57] There was some erosion in this figure over time, and continual adjustments in product coverage, exemptions for neighboring countries, allocation procedures to companies, etc., were required to shore up the program. Despite these problems, the quota was maintained more or less intact with imports increasing gradually over time and reaching a level of 23 percent of total domestic supply in 1972.[58]

By April 1972, the market for domestically produced oil was clearing at the then prevailing price, and the state authorities were permitting production at capacity. The Nixon administration adopted a deliberate policy of increasing the quota to permit imports to meet increases in demand [59] as an alternative to permitting the domestic price to rise. At the same time, an effort was made to prevent imports from inducing cutbacks of domestic production by state authorities.[60]

In April 1973 import quotas were discontinued and a tariff was introduced. The tariff was initially set at a low level, but it has recently (February and May 1975) been increased, and plans for further increases have been announced.

For purposes of this study, the primary significance of U.S. import policy is that it effectively segmented the U.S. market from the world market permitting a divergence of the U.S. price from the world price for a period of several years.

Tax Policies.[61] The petroleum industry has long received specialized treatment under U.S. tax law. The special provisions affect what items may be charged against income for tax purposes and the timing of those charges. Until March 1975 the oil producing firm had two options for depreciation of oil producing properties. Under cost depletion, which is still available, the taxpayer may deduct a proportion of the costs of a producing property equal to the units sold during the year as a proportion of the units estimated to remain in the property at the beginning of the year. Using this method, depreciation of the asset for tax purposes occurs at the same rate at which the asset is actually used up if there are no errors in estimating the resource base. Capital costs incurred in exploration for, or development of, a property which are expensed as incurred under the intangibles deduction (explained below) may not be included in the basis for cost depletion. Cost depletion deductions cease when the cost of the property has been completely deducted.

Under the depletion allowance option, the taxpayer was permitted to deduct 22 percent (27.5 percent prior to 1969) of the gross value of production (i.e., sales) from the property as long as such charges did not exceed 50 percent of net income from the property in the tax year. This option was not affected by the expensing of intangibles or the original cost of the property.

In March 1975 major changes were made in the depletion allowance provisions.[62] The major integrated firms are no longer permitted to use the depletion allowance. Small independent firms are permitted to continue using the depletion allowance, but no more than 2,000 barrels per day can qualify for the allowance. This limit is to be gradually reduced to 1,000 barrels per day by 1980. In addition, the current 22 percent rate will be gradually reduced to 15 percent by 1984. Claims for depletion will not be permitted to exceed 65 percent of a firm's taxable income.

The intangibles deduction allows the taxpayer to charge against income from any source all intangible costs of drilling and equipping productive wells. Intangibles are expenditures for non-salvageable items and are defined to include expenses for labor, power and fuel, materials and supplies, tool rental, repairs of drilling equipment, and non-recoverable materials used in drilling. There is no limit on the amount of such deductions. All exploration costs for dry holes may be expensed as incurred except geological and geophysical expenditures resulting in the acquisition or retention of properties. The latter costs may be capitalized, but cannot be deducted through cost depletion if percentage depletion is being used.[63]

A taxpayer using percentage depletion must deduct all depletion actually taken from the cost of the property in computing his tax obligation if he switches to cost depletion or sells the property.[64] In the past, the buyer could use either percentage depletion, or cost depletion based on the purchase price, regardless of the amount of percentage depletion which was taken by the previous owner.

Individual properties may not be pooled for calculating operating costs or for calculating depletion. However, net income after depletion may be pooled, and all other costs of doing business, including costs of dry holes not associated with producing properties, may be deducted prior to computation of tax.

It follows that high expenses associated with individual producing properties may not have reduced taxable income because, when percentage depletion was added, the 50 percent net income limit may have been reached. That is, the properties with high producing expenses had the smallest net incomes, and the tax laws specified that the deduction for depletion could not exceed 50 percent of net income on a property. As a result of the 50 percent limit, McDonald indicates that, for the years 1950–52, actual depletion was 25.3 percent when the statutory rate was 27.5 percent.[65] Other costs of doing business including exploratory costs, dry holes, etc., not associated with producing properties reduce taxable income until the zero income level is reached.[66]

Considered in isolation, expensing of intangibles is qualitatively similar to accelerated depreciation. The cost of the investment is written off immediately rather than being capitalized and depreciated over the life of the investment. In a firm with rapidly growing expenditures, such a provision could result in a substantial tax saving when the time stream of payments is considered. The provision allowing costs of dry holes to be written off has the same effect but does not differ from the treatment of similar expenses, such as research, in other industries.[67]

Since the majority of investment expenditures are expensed under the intangibles provision, the percentage depletion allowance effectively allowed the investment to be written off a second time. In a study for the Treasury Department, the CONSAD Corporation estimated that deductions with the depletion allowance were almost twice the amount which could have been deducted by amortizing costs which were not expensed.[68] Likewise, McDonald reports estimates based on data for 1950–52 which indicate that: "Of allowable depletion, 95.7 percent, on an average, was excess over cost basis depletion."[69] This is the basis for the frequent assertion that the depletion allowance amounted to a subsidy for the petroleum industry.

One may question whether the owners of petroleum firms and producing properties were particularly deserving of governmental largesse, but the question is one of equity which economic tools are ill equipped to solve. These tax policies were, however, roundly criticized by economists on efficiency grounds. It was argued that these tax provisions caused over-investment in the petroleum industry relative to other industries and distorted intertemporal production decisions, shifting production toward the present.[70]

Leasing of Federal Lands. The public lands of most importance from the standpoint of oil and gas exploration are offshore areas. The federal government claims jurisdiction over areas of the outer continental shelf extending beyond three miles from shore. These claims have recently been reaffirmed by the Supreme Court in a unanimous decision upholding federal claims to jurisdiction beyond the three-mile limit for the entire East Coast.[71] The only exceptions to the three-mile limit are the Gulf Coasts of Florida and Texas where state control extends nine nautical miles from shore.[72]

There appear to be no firmly established policies for determining what amounts or areas of land are to be auctioned for exploration in any given period. As table 2–4 shows, the amounts of new acreage leased annually have followed a very erratic pattern. Acreage is not sold either because no bids for the land are forthcoming,[e] or because the Secretary of the Interior chooses to reject the highest bid offered. An average of 5 percent of high bids have been rejected without explanation.[73]

While data for offshore discoveries are limited, some inferences about the relative importance of offshore lands can be made using available data. Table 2-5 contains estimates of the proportion of new oil discoveries which are from the Gulf of Mexico. This overstates the importance of federal offshore lands since a substantial portion of discoveries in the Gulf are on state lands. Federal offshore lands were not opened for exploration until 1954. If discoveries on state lands continued to be the same proportion of the U.S. total after 1954 as before, the data in table 2-5 imply that federal offshore lands contributed, on average, between 5 and 10 percent of total discoveries in the 1955-69 period. If the state share declined, the federal share would, of course, be correspondingly larger. As indicated earlier in this chapter, discovery estimates during the first few years after discovery are subject to considerable error. The figures for 1970-72 should be considered in that light.

Production from offshore discoveries has been slow in developing because of the time required to install producing wells and the necessary pipeline facilities. However, the production data in figure 2-2 are at least consistent with the estimate that offshore lands are contributing between 5 and 10 percent of new discoveries. If that estimate is correct, the curves in figure 2-2 can be expected to level off unless substantial additional offshore acreage is leased.

In the empirical analysis of this study no distinction is made between discoveries on federal and non-federal lands. The reason is that, with currently available data, it is not possible to do otherwise. The information presented above suggests that this is not likely to result in significant errors since discoveries on federal lands are a modest proportion of the total.[f] This leaves open the question of what discoveries would have been if the federal government had adopted different leasing policies.

Control of Natural Gas Prices. The final area of major federal government intervention in the petroleum market is regulation of natural gas prices by the Federal Power Commission (FPC). This policy, intended to help the consumer, was begun in 1954. The FPC has controlled prices effectively, and the result appears to have been creation of substantial excess demand as one would expect

---

[e]This may occur if prospecting firms feel that the 16 2/3 percent royalty will render exploration unprofitable.

[f]No distinction between federal and other lands would be necessary if federal leasing decisions were made in the same manner as leasing decisions of private landowners responding to market determined rents. Time does not permit a detailed study of federal leasing decisions here. It seems doubtful that the erratic pattern of federal land leasing is a response to market incentives, however.

Table 2-4.   Acreage Leased Annually: Outer Continental Shelf

Date	Acres Offered	Acres Sold
1954	860	462
1955	674	403
1956	0	0
1957	0	0
1958	0	0
1959	458	132
1960	1,611	704
1961	0	0
1962	3,679	1,909
1963	670	313
1964	1,090	580
1965	0	0
1966	0	0
1967	971	744
1968	1,270	904
1969	0	0
1970	593	544
1971	0	0
1972	971	826

Source: U.S. Congress, Senate, Committee on Interior and Insular Affairs. *Natural Gas Policy Issues and Options,* 93d Congress, 1st session, Serial No. 93–20 (Washington, D.C.: Government Printing Office, 1973), p. 53.

Table 2-5.   Gulf Coast Oil Discoveries as a Percentage of U.S. Total

1947–49	7.91
1950–54	7.78
1955–59	13.63
1960–64	14.97
1965–69	18.00
1970–72	8.55

Source: Calculated from *Reserves of Crude Oil, Natural Gas Liquids, and Natural Gas in the United States and Canada and United States Productive Capacity as of December 31, 1972* (Washington: The American Petroleum Institute, 1973), pp. 25 and 71. Alaskan discoveries are excluded in these calculations.

if prices are held below their equilibrium level in a competitive industry.[74] For purposes of the present study direct price regulation causes no problems. When price is held consistently below its equilibrium level, the supply curve can be observed. Problems are created, however, for studies seeking to estimate demand relationships.

INDUSTRY STRUCTURE AND BEHAVIOR
OF FIRMS

Analysis of this subject could easily and usefully be extended to a book-length study. Neither time nor space permit such an extensive treatment here, but behavior is a crucial element in the formulation of a model of supply. The

Figure 2-2. Outer Continental Shelf Production as Percentage of U.S. Total

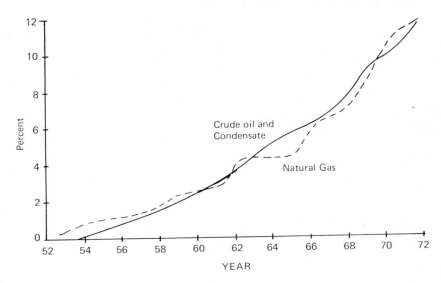

assumption of competitive behavior is utilized in the analysis of Chapter 3 and in the development of the empirical model in Chapter 5. While no amount of analysis can prove or disprove the assumption, a review of the evidence will indicate its plausibility. With this limited objective in mind, data on horizontal and vertical structure of the industry are presented below.

Structure

The broad outlines of the structure of the crude oil industry are indicated in table 2-6. In this table the shares of each of the major firms in U.S. crude oil production and refinery capacity are presented. As a rough indicator of the degree of vertical integration, the ratio of each firm's crude oil production to its domestic refining capacity is presented. Finally, as an indicator of absolute size, each firm's rank on the Fortune 500 list of the largest U.S. industrial corporations is presented.

Market concentration in the petroleum industry measured either by shares of crude oil production or shares of refinery capacity is not inordinately high. The largest four firms control just over 31 percent in both categories while the largest eight control 70 percent of crude oil production and 83 percent of refining capacity. For whatever reason, a high degree of vertical integration is an accomplished fact in the petroleum industry. The self-sufficiency ratios in table 2-6 understate slightly the actual degree of vertical integration, since refining capacity will generally not be fully utilized.

If size facilitates accessibility to capital, firms in the petroleum industry should receive very favorable terms. This will confer little relative advantage,

Table 2-6.   Market Shares and Self-Sufficiency Ratios of the Twenty
Largest Integrated Oil Companies in 1969

Company	Percentage Crude Oil Production	Percentage Refining Capacity	Percentage Self-Sufficient	Fortune 500 Rank (1970)
Standard Oil (N.J.)	9.4	8.5	87.3	1
Texaco	9.4	8.0	93.2	3
Gulf	6.5	5.9	87.5	5
Mobil	3.8	7.2	42.2	7
Standard Oil (Calif.)	5.1	5.9	68.9	10
Standard Oil (Ind.)	4.9	7.7	50.5	13
Shell	5.9	7.5	63.0	15
Atlantic-Richfield	4.9	5.7	69.0	16
Phillips	2.9	3.1	74.6	23
Continental	1.8	2.3	63.9	25
Sun	3.2	3.7	68.0	29
Union	3.1	3.4	73.4	34
Cities Service	2.1	2.0	84.2	37
Getty	3.2	1.7	153.3	45
Standard Oil (Ohio)	.3	3.0	7.9	56
Marathon	1.7	1.4	94.5	79
Amerada Hess	1.0	2.5	31.5	96
Ashland	.15	2.2	5.4	111
Kerr McGee	.33	.4	64.9	141
Skelly	.9	.6	130.9	-
Sum of Largest Four	31.2	31.7	—	—

Sources: Crude oil production and refining capacity by firm taken from Fred C. Allvine and James M. Patterson, *Competition Ltd.: The Marketing of Gasoline* (Bloomington: Indiana University Press, 1972), p. 213. Total U.S. refining capacity in 1969 was 11,575,000 barrels per day but crude oil production was only 9,238,000 barrels per day. Thus a firm can have a higher share of domestic production than of domestic refining capacity and still not be self-sufficient. U.S. refining capacity reported in *Petroleum Facts and Figures, 1971,* p. 135. Daily crude oil production reported in *The Oil Producing Industry in Your State,* 1973, p. 93. Asset rankings from *The Fortune 500 Double Directory, 1972.*

however, since nineteen of twenty firms listed in table 2-6 are within the largest 150 U.S. corporations.

Since the econometric analysis in this study uses data for 1947-68, it is of interest to observe the trends in concentration and integration during this period. Table 2-7 contains the necessary data for comparison with table 2-6.

The degree of concentration in refining has not changed in the period from 1952-68, but concentration in crude oil production has increased. It appears from the higher self-sufficiency ratios in 1969 relative to 1952 that firms have expanded their shares of crude oil production during this period to match their refinery shares.

It is also important to know the relative control of entry exercised by the largest oil producing firms. A key element of such control would reside in control of access to oil-bearing land. Unfortunately, recent data on the distribution of lease holdings are not available. The information in table 2-8,

Table 2–7. Market Shares and Self-Sufficiency Ratios of the Twenty Largest Integrated Oil Companies in 1952

Company	Percentage Crude Oil Production	Percentage Refining Capacity	Percentage Self-Sufficient
Standard Oil (N.J.)	6.5	11.2	52.1
Socony-Vacuum	3.5	7.3	43.1
Standard Oil (Ind.)	3.9	7.3	48.4
Texas Company	4.8	6.8	63.9
Gulf	3.6	6.5	49.3
Standard Oil (Calif.)	4.1	5.8	64.2
Cities Service	1.3	3.2	38.1
Sinclair	1.7	5.2	30.4
Shell	4.2	5.5	68.8
Phillips	2.0	3.1	56.7
Atlantic Refining Co.	1.4	2.5	51.3
Union Oil	1.5	2.0	64.3
Sun Oil	1.7	3.1	49.5
Pure Oil	1.1	1.6	64.1
Continental	1.8	1.6	103.0
Tidewater Associated	1.5	2.1	63.7
Standard Oil (Ohio)	.5	1.7	27.1
Ohio Oil Co.	1.5	.5	244.8
Skelly	1.0	.6	150.3
Richfield	.9	1.5	55.1
Sum of Largest Four	19.6	32.6	—

Source: John G. McLean and Robert Wm. Haigh, *The Growth of the Integrated Oil Companies* (Boston: Division of Research, Harvard University, Graduate School of Business Administration, 1954), p. 332.

though somewhat dated, is still of considerable interest. In 1956, the largest four firms held 18 percent of total acreage, and the largest eight held 32 percent. The share of the largest four firms in total acreage in 1956 was approximately the same as their share of production in 1952.

By these figures, the large firms were far from having a stranglehold on leases in 1956. Of course, quality of land under lease in terms of oil-bearing potential is more important than total acreage, but no measure of the quality of land under lease is available. In reviewing these data, McKie concludes, "Thus the major companies account for a larger share of lease acreage and scientific exploration than of drilling, *though the large firms individually and collectively do not control access to unleased land* or the supply of exploratory and drilling facilities." [75] (emphasis added)

Implications for Behavior

While the evidence is inconclusive, it does indicate that control of crude oil production is distributed among several firms, none of which holds a dominant position. Collective action by firms to control production and price cannot be

Table 2-8.   Undeveloped Acreage Under Lease to Twenty Largest Leasees in the U.S. March 1, 1956

Company	Acreage	Percentage of Total
Shell	16,893,000	5.36
Humble	15,087,000	4.79
Standard Oil (Ind.)	13,106,000	4.16
Gulf	12,275,000	4.04
Magnolia	11,456,000	3.63
Sinclair	11,021,000	3.50
Texas Co.	10,877,000	3.45
Sun Oil Co.	9,870,000	3.13
Phillips	8,694,000	2.76
Continental	8,643,000	2.74
Standard Oil (Calif.)	8,289,000	2.63
Superior	6,670,000	2.12
Amerada	5,623,000	1.78
Pure	5,519,000	1.75
Cities Service	5,355,000	1.70
Carter Oil Co.*	5,348,000	1.70
Skelly	4,827,000	1.53
Atlantic Refining Co.	4,411,000	1.40
Union Oil	4,294,000	1.36
Ohio Oil Co.	3,604,000	1.14
Sum of Four Largest	57,811,000	18.34
Sum of Twenty Largest	172,331,000	54.67
Total U.S.	315,211,000	100.00

Source: James W. McKie, "Market Structure and Uncertainty in Oil and Gas Exploration," *Quarterly Journal of Economics* 74, #4 (November 1960): 548.

*Carter and Humble are affiliates of Standard Oil of New Jersey.

ruled out entirely. However, evidence from the preceding sections suggests rather strongly that the states have been responsible for control of production and thereby prices, and that firms have behaved competitively within the bounds established by state laws.

If the firms were able to act cooperatively, it would have been in their interest to limit exploratory activity. Control of price could have been exercised in this manner without the loss in foregone earnings entailed in developing and maintaining the substantial excess capacity evident in tables 2-2 and 2-3. Furthermore, if firms had chosen to maintain price by allowing excess capacity to develop, it is not clear why that excess capacity would have been distributed so unevenly across states (table 2-3). The evidence suggests instead a desire by the major producing states to maintain price while affording wide opportunity for firms to "buy in" by finding and developing new reservoirs. This evidence is the basis for the assumption of competitive behavior used in developing the behavioral model in succeeding chapters.

Chapter Three

Economic Rents and the Price Elasticity of U.S. Petroleum Supply

INTRODUCTION

Of the published empirical studies of crude petroleum supply, one of the more ambitious is an attempt by Richard Manke to estimate the overall price elasticity of U.S. domestic petroleum supply using what he terms the "residual supply curve" approach.[1] The key element of this approach is the use of an estimate of the share of total revenues paid as economic rents to determine the elasticity of the supply curve. While there are shortcomings to Mancke's approach, the idea is an ingenious one.

In this chapter a two-factor model of petroleum production will be developed. The analysis of this model provides a convenient framework for reviewing Mancke's methodology. More importantly, the results of this analysis supplemented with empirical information concerning the parameters of the model can be used to estimate a range of values for the price elasticity of domestic petroleum supply.

The elasticity of supply of any good depends on the elasticity of supply of the inputs used to produce it and the technological opportunities for substituting among them. In the case of crude oil in the United States, capital, labor, and materials may be assumed to be in highly elastic supply to the industry except in the relatively short run. Oil-bearing land, however, is not available in perfectly elastic supply. Mancke has argued that the elasticity of supply of crude oil can be inferred from an analysis of the share of crude oil revenues which are paid as economic rents for the use of oil-bearing land. In using this so-called "residual supply curve approach," however, he fails to take account of the opportunities for substituting other inputs for oil-bearing land. His procedure thus yields erroneous results as the following analysis will demonstrate.

THE RESIDUAL SUPPLY CURVE APPROACH

Mancke prefaces his analysis with the observation that discounted expected revenues equal discounted expected real (i.e., non-rent) costs plus discounted expected rents. He then describes his procedure as follows:

First, draw in price-quantity space a rectangle whose area measures the crude oil industry's expected total revenue over some time span. The length of the vertical axis of this rectangle is equal to crude oil's expected price. The length of the horizontal axis of this rectangle is equal to the total quantity of crude oil that the industry expects to sell at this price. Second, subtract from this crude oil's expected price all expected rents accruing to each expected barrel of crude oil output. Then the residual curve must measure the crude oil industry's *ex ante,* real, longrun marginal cost (supply) curve.[2]

After estimating the proportion of expected revenues that were lease bonuses in the period 1955 to 1967, Mancke states:

The foregoing calculation demonstrates that total lease bonuses have been a relatively small component (about 16.4 percent) of crude oil's expected net total costs. Knowledge of this constraint lets us immediately dismiss the common contention . . . that the United States' longrun 'lower 48' crude oil supply curve, over its entire range, is less than unitary price-elastic, because all such price-inelastic longrun supply curves imply that total lease bonuses must be more than 50 percent of crude oil's expected net total costs.[3]

In essence, Mancke is arguing that a necessary condition for the supply curve to be less than unitary elastic over its entire range is that economic rents exceed 50 percent of total factor payments. It will be shown below that without the qualifier "over its entire range" this statement is incorrect, and that with this qualifier the statement is technically correct but empirically uninteresting.

A TWO-FACTOR MODEL OF SUPPLY

In the following analysis it is assumed that:

1. product and factor markets are competitive,
2. the objective of oil producing firms is profit maximization,
3. the industry production function exhibits constant returns to scale, and
4. a composite good, representing all non-rent-earning inputs is available to the industry in perfectly elastic supply.

The role of each of these assumptions is made clear as the derivation proceeds.
    The production function may be written

$$Q = F(M,N)$$

$$(3-1)$$

where Q is output, M is the rent-earning input, and N is the composite good representing all other inputs. The profit from oil production may be written:

$$\pi = PQ - P_m M - P_n N \qquad (3\text{-}2)$$

where P, P_m, and P_n are the prices of Q, M, and N respectively. Maximizing (3-2) subject to the constraint imposed by (3-1) yields the following first order conditions:

$$PF_m = P_m \qquad (3\text{-}3)$$

$$PF_n = P_n \qquad (3\text{-}4)$$

where F_m and F_n are the partial derivatives of F with respect to M and N. The assumptions of competitive factor and product markets are implicit in equations (3-3) and (3-4) since factor and product prices were treated as parameters in performing the differentiation. The factor supply equations are:

$$P_n = \text{Constant} \qquad (3\text{-}5)$$

and,

$$M = AP_m^b \qquad (3\text{-}6)$$

The input M has an elasticity of supply b and may be thought of as oil-bearing land.[a] The input N represents capital, labor, and materials used by oil producers. It is assumed that the prices of these inputs are determined by the level of activity in the overall economy and thus fixed with respect to the petroleum industry.

Equation (3-7) is the condition that total revenues equal total costs in a competitive industry.

$$PQ = P_m M + P_n N \qquad (3\text{-}7)$$

Equations (3-8) through (3-10) are the definitions of factor shares and the elasticity of substitution respectively.

$$S_m = \frac{P_m M}{PQ} \qquad (3\text{-}8)$$

$$S_n = \frac{P_n N}{PQ} \qquad (3\text{-}9)$$

[a]The term oil-bearing land should be understood to include not only land on which oil has been discovered but also land which is potentially productive of oil.

$$s = \frac{-d \, \text{Log} \left(\dfrac{M}{N} \right)}{d \, \text{Log} \left(\dfrac{F_m}{F_n} \right)} \tag{3-10}$$

The objective now is to derive the elasticity of supply of output Q in terms of the parameters of the model and factor shares.

The total differential of equation (3-1) is

$$dQ = F_m dM + F_n dN \tag{3-11}$$

Substituting equations (3-3), (3-4), (3-8), and (3-9) into (3-11) and simplifying yields:

$$\frac{dQ}{Q} = S_m \frac{dM}{M} + S_n \frac{dN}{N} \tag{3-12}$$

If the production function exhibits constant retuns to scale, equation (3-7) will hold for all equilibrium values of the variables. Total differentiation of equation (3-7) yields:

$$PdQ + QdP = P_m dM + P_n dN + MdP_m + NdP_n \tag{3-13}$$

After substituting equations (3-3), (3-4), (3-11), (3-8), and (3-9) respectively into (3-13) and simplifying, one obtains:

$$\frac{dP}{P} = S_m \frac{dP_m}{P_m} + S_n \frac{dP_n}{P_n} \tag{3-14}$$

Equations (3-15) and (3-16) below are obtained by differentiating equations (3-5) and (3-6).

$$\frac{dM}{M} = b \frac{dP_m}{P_m} \tag{3-15}$$

$$\frac{dP_n}{P_n} = 0. \tag{3-16}$$

Substituting equations (3-3) and (3-4) into (3-10) and simplifying yields the following result:

$$\frac{dM}{M} - \frac{dN}{N} = s \left(\frac{dP_n}{P_n} - \frac{dP_m}{P_m} \right). \tag{3-17}$$

Finally, solving equations (3-12) through (3-17) for the elasticity of supply of Q yields:

$$e = \frac{P}{Q}\frac{dQ}{dP} = \frac{b + s\,S_n}{S_m} = \frac{b + s - s\,S_m}{S_m} \tag{3-18}$$

where the requirement that $S_m + S_n = 1$ (apparent from equations (3-8) and (3-9) has been used to obtain the final expression in equation (3-18).[b]

Equation (3-18) expresses the elasticity of supply of the product in terms of the share of income going to the rent-earning factor. In general, not all payments to factor M are rents. Only the shaded areas between the supply curve and the horizontal line at P_m in figure 3-1 should be considered rents. By integrating equation (3-6) it is found that rents are:[c]

$$R = \int_0^{P_m} A p_m{}^b\, dp_m = \frac{A P_m{}^{b+1}}{b+1} = \frac{P_m M}{b+1} \tag{3-19}$$

From equations (3-8) and (3-19), the share of total factor payments that are rents is:

$$S_r = \frac{R}{PQ} = \frac{P_m M}{(b+1)PQ} = \frac{S_m}{b+1} \tag{3-20}$$

Substituting equation (3-20) into (3-18) yields the desired result.

$$e = \frac{b+s}{(b+1)S_r} - s \tag{3-21}$$

Equation (3-21) shows that the elasticity of supply of crude oil, e, is a function of the elasticity of supply of oil-bearing land, b; the elasticity of substitution between oil-bearing land and other inputs, s; and the rent share S_r. It demonstrates that the residual supply curve approach which relies only on an estimate of S_r to determine e is inadequate. In the paragraph quoted earlier, Mancke claims that the elasticity of supply of crude oil must exceed one since

---

[b]In an earlier analysis,[4] Richard Muth derives the supply curve of a produced good under general conditions in which both factors are in less than perfectly elastic supply and shifts in the production functon and the supply curves may occur. Muth's equation (19), p. 227, can be specialized to equation (3-18) given in this chapter under the conditions assumed for the model developed here. The results derived subsequent to equation (3-18) are not available in Muth's paper and must be derived independently.

[c]Implicit in the integration is the assumption that the elasticity of supply of the rent-earning factor is constant within the relevant price range. This is qualitatively stronger than the assumption used in the remainder of the analysis in which parameters are assumed constant only for differential changes in the variables.

Figure 3-1. Economic Rents as a Proportion of Factor Payments

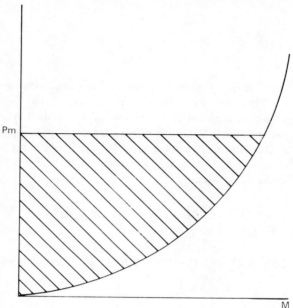

the rent share is only 16.4 percent.[5] As a simple counter-example to this claim, consider the case where b is zero. Substituting $S_r = .164$ and $b = 0$ into equation (3-21), one obtains:

$$e = s\left(\frac{1}{.164} - 1\right) \cong 5s. \tag{3-22}$$

Thus, any value for the elasticity of substitution less than 0.2 would imply an elasticity of crude oil supply which is less than one.

There are two special cases of equation (3-21) for which Mancke's assertions are true. When the elasticity of substitution, s, is one, equation (3-21) reduces to $e = \frac{1}{S_r} - 1$, and e must exceed one for any value of S_r less than one-half. Alternatively, when $s = 0$ and $b = 1$, equation (3-21) becomes $e = \frac{1}{2S_r}$, and again e must exceed one if S_r is less than one-half. Neither of these cases is at all general and neither appears to be very realistic as regards the oil industry; and there is nothing in Mancke's paper to suggest that he was assuming either case in his analysis.

Mancke does of course qualify his conclusion by saying that the supply curve cannot be less than unitary price elastic "over its entire range." It is true that, with an elasticity of substitution less than one, the share of income accruing to rents would exceed one-half if the price of crude oil were to rise sufficiently. The important point, however, is that within the range of prices observed in the 1955 to 1967 period a rent share of 16.4 percent is consistent with an elasticity of oil supply less than one, and if the elasticity of substitution were in fact equal to 0.2 with b equal to zero, the price of crude oil would have to rise to almost twice its 1967 value for the rent share to grow to one-half of total factor payments.[d]

ESTIMATES OF THE PARAMETERS OF THE MODEL

To estimate the elasticity of crude oil supply using equation (3–21), estimates of b, s, and S_r must be obtained. Available information concerning these parameters will now be discussed.

The Supply of Oil-Bearing Land

To develop an estimate of b, it is useful to consider the alternative uses for oil-bearing land. With the relatively rare exception of those situations in which oil is found in an urban area, the alternative use for oil-bearing land is agriculture. Oil production normally requires relatively modest amounts of surface land, and the foregone income from crop production is thus relatively small. Indeed, oil extraction and agriculture are quite compatible in most cases. To a close approximation, then, the reservation price of oil-bearing land is zero. An equivalent statement of this argument is that the price of oil-bearing land could fall to zero without substantially affecting the amount of such land available for oil exploration. Viewed in this light, the elasticity of supply of oil-bearing land would appear to be approximately zero.

There is, however, a degree of ambiguity in defining precisely what constitutes oil-bearing land. Oil is a non-reproducible resource. Unlike land used in agriculture, oil-bearing land is essentially used up as oil is produced. The determination of the rate at which a resource in fixed total supply will be produced and consumed in a competitive economy is an extremely difficult problem even when perfect certainty about present and future demand and cost functions is assumed.[e] It is important, however, to recognize that a resource whose supply is fixed will be used up gradually even if the cost of making it available is negligible. Owners of the rights to oil-bearing land might therefore choose to withhold their land from exploration, not because exploration

---

[d]This result is not readily apparent from the analysis leading to equation (3–21) since only differential changes are considered in that derivation. The result can be derived using a CES production function with an elasticity of substitution of 0.2 assuming N is in fixed supply and the prices of other inputs are fixed.

[e]An analysis of this problem is taken up in Chapter 5.

interferes with other productive activities, but because they expect that in the future prices will rise sufficiently to compensate them for not making the land available immdeiately. Clearly, however, the expected rate of increase in price would have to exceed the opportunity rate of return on investments, or the landowner would have an incentive to lease the land immediately and invest the proceeds.

During the period from which the data for this analysis are taken, the price of crude oil changed very little. As a consequence, the rentals offered for oil-bearing land probably increased very little if at all, and certainly less than the rate of return which landowners could have earned on invested income.[f] In short, the expectation of future rental increases had an insignificant effect on the supply of oil-bearing land during the period in question.

Finally, there is the question of the quality of oil-bearing land in use. In the analysis leading to equation (3-21), the production function was assumed to exhibit constant returns to scale. Thus, in this analysis, any diminution in the quality of land is reflected in the shape of the supply curve for oil-bearing land and not in diminishing returns in the production function. Even if a change in price did not affect the willingness of landowners to make land available, that price increase might result in use of land which was previously considered to be unprofitable. This will impart a positive slope to the supply curve for oil-bearing land.

How much is the quantity of land in use, measured in homogeneous units, affected by a change in the price of crude oil? Since no suitable measure of oil-bearing land is available, the question cannot be given a precise answer. Casual observation suggests, however, that an increase in the price of oil does result in drilling in areas of the United States which had previously not been extensively explored, but that discoveries in such areas contribute very little to the total quantity of oil discovered. This in turn implies that quality variations impart a modest positive elasticity to the supply curve. An upper bound estimate of the elasticity of supply of oil-bearing land of one-half will be considered in subsequent calculations. While it is doubtful that quality variations in land would result in an elasticity that high, the calculations will serve to indicate the sensitivity of the results to changes in the value of b.

The Elasticity of Substitution

The elasticity of substitution between oil-bearing land and other inputs is by far the most difficult parameter to estimate. A rough estimate of the range of values which s may assume can be made, however. The elasticity of substitution is a measure of the curvature of a production isoquant. It is thus a measure of

---

[f]Changes in the unit rental on oil-bearing land are not necessarily of the same order of magnitude as changes in the price of oil. The relative stability of the royalty share on privately owned land during this period suggests, however, that land rentals were not increasing at a more rapid rate than oil prices.

the extent to which the input ratio changes in response to a change in the relative prices of the inputs. An infinite elasticity of substitution implies a straight line isoquant while a zero elasticity of substitution corresponds to the case of fixed coefficients for which the isoquants are L-shaped. It is clear that the elasticity of substitution of oil-bearing land for other inputs is greater than zero. Other inputs may be substituted for oil-bearing land by more intensive geological and geophysical prospecting, by drilling more and deeper exploratory wells, and by use of more costly recovery methods to obtain additional oil from known reservoirs.

An effort to pin down an estimate of the elasticity of substitution in oil production would be made considerably easier if there were a large body of evidence on the elasticity of substitution among various inputs in other industries. Unfortunately, empirical work to date has concentrated almost exclusively on obtaining estimates of the elasticity of substitution between labor and capital in a variety of industries. Nerlove tabulated the results of a number of these studies in his 1967 survey of empirical evidence on the CES production function.[6] He finds distressing divergences in estimates of the elasticity of substitution for particular industries by researchers using similar data and techniques. However, a review of the tables which Nerlove presents reveals that published estimates of the elasticity of substitution between labor and capital on the order of one are the rule with estimates below one being somewhat more common than estimates exceeding one. In the analysis cited earlier, Muth estimates that the elasticity of substitution between land and non-land inputs in the production of housing is about 0.75.[7] While there are opportunities for substituting other inputs for oil-bearing land in producing oil, those opportunities appear to be more limited than the opportunities for substituting labor for capital in most industries and no greater than the opportunities for substituting land for non-land inputs in the production of housing. A rough guess is that the elasticity of substitution of oil-bearing land for other inputs is on the order of 0.5, but values near zero or as high as one cannot be ruled out a priori.

The Rent Share

In his analysis Mancke develops annual estimates of the rent share in oil production for the 1955 to 1967 period. His estimation procedure is briefly reviewed below, and some modifications in his estimates are suggested. The theoretical justification for his procedure is that, in a competitive industry, expected future revenues will equal expected future costs. Current economic rents divided by expected revenues gives an estimate of the share of total costs which are economic rents. The procedure thus requires that he first estimate annual petroleum industry expenditures for items which are in the nature of economic rents. The next step is to obtain an estimate of the present value of expected revenues from oil production since expected revenues are assumed to be equal to expected costs. The ratio of rent payments to the present value of expected revenues then serves as an estimate of the rent share.

Mancke treats lease bonuses and lease rentals as economic rents. He observes that royalty payments are also a form of economic rent, but, rather than include these payments with lease bonuses and lease rentals, he reduces the wellhead price of crude oil by about 15 percent to account for royalty payments.[8] If the theoretically more appealing alternative of adding these payments to lease bonuses is adopted, the estimated share accruing to rents increases from 16.4 percent to about 25 percent.[g]

If revenues from the production of oil from a reservoir were obtained at a single point in time, discounting those revenues to the present would be a trivial problem. Instead, reservoirs produce for several years, and the discounting procedure must account for the time distribution of the receipt of revenues from the recovery of oil. Mancke assumes an exponential production decline to develop an adjustment factor to discount expected revenues back to the point at which economic rents are paid. Evidence in the preceding chapter indicated that the exponential decline does provide a reasonable approximation of reservoir production patterns.

In estimating expected future income Mancke postulates that the oil producer assumes that price will remain unchanged, but that the quantity of output will continue to grow at the rate observed at the time the estimate of future revenues is made.[9] He fails to explain why supply will continue to grow in the absence of a price change. A repeat of Mancke's calculations after elimination of the assumed growth in output in the absence of a price change reveals that the estimate of the share of income accruing to rents should be increased by about 15 percent. Applying this adjustment to the figure obtained above yields a new estimate of the rent share of about 0.3.

Finally, in developing an adjustment factor to discount expected revenues back to the date at which lease bonuses and rentals are paid, Mancke assumes that Moody's Baa bond rate is a reasonable estimate of the after tax cost of capital in oil exploration. To justify the use of this discount rate, he states that the marginal tax rate of firms engaged in oil exploration is 50 percent and that the real cost of debt capital to those firms is therefore half the market rate.[10] That the marginal tax rate of firms engaged in oil exploration is as high as 50 percent is certainly debatable. It is well known that firms in the oil industry have enjoyed special tax provisions which allowed them to deduct all expenses for dry holes plus intangible expenses for successful wells plus a percentage of revenue from oil production. After taking advantage of these deductions, many firms engaged in crude oil exploration may have had more than enough tax deductions to offset their entire tax liability. If that is the case, the ability to deduct interest

---

[g]The share accruing to rents is defined as total rent payments divided by total revenues from oil production. The change suggested above involves adding .15 times total revenues (as estimated by Mancke) to both the numerator and denominator of this fraction.

on debt capital would not have reduced the real cost of that capital to such firms. Unfortunately, little information is available which might resolve this question, but some upward adjustment of Mancke's estimate of the discount rate is probably warranted because of these tax considerations.

Anticipating skepticism about his choice of discount rate, Mancke argues that his adjustment factor could be no more than doubled as a result of revising the stimated discount rate.[11] If the adjustment factor were doubled, calculations using Mancke's data reveal that the estimate of the rent share rises to about 45 percent when modifications suggested in the previous paragraph are incorporated.[h] This may then be taken as an upper bound estimate of the share of income accruing to rents.

SOME ESTIMATES OF THE ELASTICITY OF CRUDE OIL SUPPLY

Estimates of the elasticity of crude oil supply using the range of parameter values suggested in the preceding discussion are tabulated in table 3-1. These values are calculated from equation (3-21). On the basis of the preceding discussion, $b = 0$, $s = .5$, and $S_r = .35$ are the most plausible estimates of the parameter values. As the table shows, these values imply an elasticity of crude oil supply of 0.93. If the parameter values in this table bracket the true values—and they almost surely do—the minimum and maximum values for the price elasticity of crude oil supply are zero and 2.3 respectively. Any narrowing of the range of values for the underlying parameters involves judgments which are not easily verified. In any case, the range of values in table 3-1 are certainly not consistent with Mancke's claim that the supply curve for crude oil is highly elastic. It is very doubtful that the true parameter values lie outside the range of values in that table.

Table 3-1.   Estimates of The Elasticity of Crude Oil Supply

	$b = 0$			$b = .5$		
S_r	.30	.35	.45	.30	.35	.45
s						
0	0	0	0	1.1	.95	.74
.5	1.2	.93	.61	1.7	1.4	.98
1.0	2.3	1.9	1.2	2.3	1.9	1.2

The Elasticity of Substitution Reconsidered

As noted earlier, the most difficult parameter to estimate is the elasticity of substitution. It is well known that the share of income accruing to each factor of production will be constant if the elasticity of substitution is equal to one (i.e.,

[h]The adjustment factor is not applied to royalty payments in either case since such payments are not made until production is realized.

if the production function is Cobb-Douglas). On the basis of Mancke's finding that the rent share was approximately constant over the period from 1955 to 1967, it is tempting to conclude that the elasticity of substitution equals one.[i] Unfortunately, this conclusion is not justified because of shifts in the underlying elements of the crude oil supply function during the period in question. To demonstrate this point, the model used previously will be modified to allow for technical progress and shifts in the input supply curves.

The production function may be rewritten as:

$$Q = F(M,N,t) \tag{3-23}$$

where t is included to allow for technical progress in production. The factor supply equations may be rewritten as:

$$P_n = P_o e^{\beta t} \tag{3-24}$$

and

$$M = AP_m{}^b e^{\gamma t} \tag{3-25}$$

where P_o is the price of N at time zero, β represents exogenous changes in the price of non-land inputs, and γ represents shifts in the supply curve for oil-bearing land.[j] The previous equations (3-3), (3-4), and (3-7) through (3-10) are unchanged.

The following equation, obtained by differentiating equation (3-23) and substituting from equations (3-3), (3-4), (3-8), and (3-9) replaces equation (3-12).

$$\frac{1}{Q}\frac{dQ}{dt} = \frac{S_m}{M}\frac{dM}{dt} + \frac{S_n}{N}\frac{dN}{dt} + \theta \tag{3-26}$$

In this equation $\theta = \dfrac{F_t}{Q}$ is the rate of technical progress which, for simplicity, is assumed to be neutral with respect to M and N. The equation below is obtained by following the same sequence of steps previously used to obtain equation (3-14).

$$\frac{1}{P}\frac{dP}{dt} = \frac{S_m}{P_m}\frac{dP_m}{dt} + \frac{S_n}{P_n}\frac{dP_n}{dt} - \theta \tag{3-27}$$

---

[i] Mancke does not make this claim. The analysis presented here is intended to demonstrate why constancy of the rent share was not used in the earlier attempt to obtain an estimate of the elasticity of substitution.

[j] It will be argued later that there is reason to believe that both of these factor supply equations did shift during the period in question.

Differentiation of equations (3-24) and (3-25) yields:

$$\frac{1}{P_n}\frac{dP_n}{dt} = \beta \qquad\qquad (3\text{-}28)$$

and

$$\frac{1}{M}\frac{dM}{dt} = \frac{b}{P_m}\frac{dP_m}{dt} + \gamma. \qquad\qquad (3\text{-}29)$$

Substituting equation (3-8) into (3-20) and differentiating yields:

$$\frac{1}{S_r}\frac{dS_r}{dt} = \left(\frac{1}{b+1}\right)\left(\frac{1}{P_m}\frac{dP_m}{dt} + \frac{1}{M}\frac{dM}{dt} - \frac{1}{P}\frac{dP}{dt} - \frac{dQ}{dt}\right). \qquad\qquad (3\text{-}30)$$

Equation (3-17) may be rewritten as:

$$\frac{1}{M}\frac{dM}{dt} - \frac{1}{N}\frac{dN}{dt} = s\left(\frac{1}{P_n}\frac{dP_n}{dt} - \frac{1}{P_m}\frac{dP_m}{dt}\right). \qquad\qquad (3\text{-}31)$$

Equations (3-26) through (3-31) may now be used to derive the following result:

$$\frac{1}{S_r}\frac{dS_r}{dt} = [-1 + \frac{1}{(b+1)S_r}]\,(\frac{1}{P}\frac{dP}{dt} + \theta - \beta)\,(1 - s). \qquad\qquad (3\text{-}32)$$

This equation expresses the time rate of change of the rent share in terms of the parameters of the model and the time rate of change of the price of crude oil. The first point to note is that γ does not appear in the above expression. Thus, shifts in the supply curve for oil-bearing land of the form assumed in equation (3-25) do not *directly* affect the share of income accruing to rents. This is a particularly useful result since there is evidence that the supply curve for oil-bearing land has shifted,[k] but measuring the magnitude of those shifts would be very difficult.

This does not mean that a shift in the supply curve for oil-bearing land has no effect on the share of income accruing to rents. An outward shift in the supply curve for oil-bearing land will *indirectly* affect the share accruing to rents by

[k]The evidence is the following. There are large year-to-year fluctuations in government lease sales, and lease bonuses on such sales are typically quite high. If the supply curve for oil-bearing land did not shift, one would expect that relatively marginal land would be brought under lease each year with correspondingly modest lease bonuses being paid for such land. The observed conditions are thus more in accord with a shifting of the supply curve for oil-bearing land than with a movement along that curve.

inducing a fall in the price of oil-bearing land which, in turn, will induce a fall in the price of crude oil. This fall in the price of crude oil will then induce a change in the rent share as shown in equation (3–32).

Mancke's data suggest that the change in rents during the period 1955 to 1967 was approximately zero. Since the first term on the right hand side of equation (3–32) is clearly not equal to zero, the observation that $\dfrac{dS_r}{dt}$ is approximately zero implies that:

$$\left(\frac{1}{P}\frac{dP}{dt} + \theta - \beta\right)(1 - s) \cong 0. \tag{3-33}$$

Constancy of the share of income accruing to rents thus implies that either the elasticity of substitution is equal to one, or that the first expression in equation (3–33) is equal to zero. The terms in that expression will now be evaluated.

The average increase in crude oil prices from 1955 to 1967 was .7 percent per year.[1] Thus $\dfrac{1}{P}\dfrac{dP}{dt} \cong .007$. There is no series which can be used directly as an index of the costs of non-land inputs in crude oil production. However, from 1955 to 1967, hourly earnings in crude oil production increased at an average annual rate of 3.1 percent, oil field machinery prices increased at an average annual rate of 1.5 percent, and oil well casing prices increased at an average annual rate of 2.3 percent.[13][m] These figures suggest that β was in the range of .02 to .03 during the period in question. Estimates of technical progress have not been made for crude oil production, and considerable controversy surrounds estimates of technical progress which have been made for the U.S. economy. There does, however, appear to be agreement that productivity growth in the U.S. economy was in the range of 1 and 1.5 percent per year from 1950 to 1962.[14] Productivity growth in crude oil production was probably higher than this because the estimate for the U.S. economy includes the service sector which has a relatively low rate of productivity growth. A figure on the order of 2 percent per year is thus more plausible for crude oil production. The above discussion thus suggests that $\dfrac{1}{P}\dfrac{dP}{dt} \cong .007$, $\beta \cong .025$, and $\theta \cong .02$. These figures clearly are intended to be only rough estimates.

Substituting them into equation (3–33) reveals that the first term of that expression was near zero during the period in question. It is thus quite possible that the share of payments going to rents did not change during this period because growth in productivity coupled with the modest rise in the price of crude oil approximately offset the increase in the cost of non-land inputs. If this

[1]Calculated from data tabulated by Mancke.[12]

[m]All figures reported above are geometric averages to conform with the assumptions made in deriving equation (3–32).

was the case in the past, there is clearly no reason why it should continue to be the case in the future. If the recent rapid rises in the price of crude oil continue, equation (3-32) would lead one to expect a rise in the share of income going to rents if the elasticity of substitution is less than one. If this does not occur, the alternative hypothesis that the elasticity of substitution is approximately one will be supported. This in turn would imply a relatively high price elasticity for the supply curve for crude oil as the figures in table 3-1 indicate.

CONCLUSION

Knowledge of the proportion of revenues which are economic rents is not alone sufficient for a determination of the price elasticity of U.S. petroleum supply. This information is valuable, however, when used in conjunction with estimates of the underlying parameters of a model of petroleum production. The viability of this approach is dependent on the reliability of the estimates of the parameters of the model. Due primarily to uncertainty about the value of the elasticity of substitution, the estimated range of values which the elasticity of petroleum supply might assume is rather large. The analysis suggests a range from 0 to 2.3 with values near the middle part of that range being most plausible.

A difficulty inherent in the approach of this chapter is that probabilistic statements about the accuracy of the parameter estimates cannot be made. That, however, is attributable to scarcity of data on which to base parameter estimates rather than to any weaknesses inherent in the theoretical model. The intended contribution of the analysis of this chapter is in identifying the underlying elements of the supply function for crude oil, and in demonstrating how those elements interact to determine the price elasticity of crude oil supply.

Chapter Four

A Critique of Prior Econometric Studies

INTRODUCTION

Several econometric studies of various facets of domestic petroleum supply have appeared in recent years. Aspects of these studies which receive particular attention in the following review are the economic foundations of the models, consistency of methodology with the assumptions of the underlying models, long-run properties of the models, and techniques used in empirical estimation. These criteria will not be used in an attempt to rank the studies. Instead, they will serve as a point of departure in assessing the strengths and weaknesses of the models.

THE CONSAD STUDY

The CONSAD Research Corporation was commissioned by the U.S. Treasury Department to perform a study of *The Economic Factors Affecting the Level of Domestic Petroleum Reserves.* [1] The study, published in 1969, concentrates particular attention on determining the effect on reserves of a change in the oil depletion allowance. CONSAD uses three alternative procedures in the analysis, but greatest reliance is placed on what is termed a "Reserve Reaction Forecasting Model." This approach is an adaptation of the methodology developed by Jorgenson in his studies of investment behavior.[2] The application of this technique to petroleum supply is based on the argument that reserves may be treated as a capital stock used in crude oil production.

At the outset of the analysis they assume that any tax increase resulting from a change in the depletion allowance is "absorbed by petroleum producers and *not* passed forward to consumers or backward to landowners." (Emphasis in original.)[3] This assumption is clearly inconsistent with competitive profit maximization under equilibrium conditions. A competitive equilibrium implies that only normal profits are being earned by firms in the industry. A tax increase would result in some or all firms earning below normal profits in the short run. A new equilibrium would not be achieved until the normal profit position was restored by passing the tax increase forward to consumers and/or backward to input suppliers.[a] Adjustment to a new equilibrium is, however, explicitly assumed in the CONSAD study as the following statement indicates: "This approach assumes that the observed values of variables are those existing under a close approximation of economic equilibrium and, consequently, that the projected levels of reserves represent those which would exist after the industry had made a full adjustment to the tax change."[4] The study thus rests on a set of logically inconsistent assumptions.

The CONSAD report states that this assumption will give an estimate of the maximum reserve impact.[5] This contention can be analyzed with the model of the preceding chapter. Redefine M to be petroleum reserves and N to be non-reserve inputs. Suppose, for the sake of argument, that petroleum producers can be made to bear the full cost of a reduction in the depletion allowance without a change in either the price or quantity of crude oil produced. Rewrite Equation (3-6) of Chapter 3 as follows:

$$M = A(uP_m)^b \qquad (4\text{-}1)$$

where P_m is the price per unit of reserves in the absence of the depletion allowance, and u is the percentage increase in this price resulting from the depletion allowance. Equation (3-15) of Chapter 3 becomes:

$$\frac{dM}{M} = b\frac{dP_m}{P_m} + b\frac{du}{u} \qquad (4\text{-}2)$$

and the total differential of Equation (3-2), Chapter 3, yields:

$$d\pi = -P_m \, dM - P_n \, dN - M \, d\,P_m - N \, d\,P_n \qquad (4\text{-}3)$$

The requirements that

$$d P = 0 \qquad (4\text{-}4)$$

and

$$d Q = 0 \qquad (4\text{-}5)$$

were utilized in obtaining Equation (4-3). With these alterations of the model of the preceding chapter, the following results can be derived:

---

[a]If inputs were in perfectly inelastic supply, the tax increase would be passed backward to input suppliers. If inputs were in perfectly elastic supply, the tax increase would be passed forward to consumers.

$$\frac{dM}{M} = \left(\frac{b \, s \, S_n}{b+s \, S_n}\right)\frac{du}{u} \, , \tag{4-6}$$

and

$$d\pi = P \, Q \left(\frac{b \, S_m}{b+s \, S_n}\right)\frac{du}{u} \, . \tag{4-7}$$

The assumption that changes in taxes will not be passed back to landowners is justified in a competitive model only if reserves are in perfectly elastic supply. Utilizing this assumption, the following results can be obtained by taking limits as b approaches infinity in the above equations:

$$\frac{dM}{M} = s \, S_n \frac{du}{u} \, , \tag{4-8}$$

and

$$d\pi = P_m \, M \frac{du}{u} \tag{4-9}$$

If a reduction in the depletion allowance is being considered, du/u will be negative. Under the CONSAD assumptions, Equation (4-8) shows that the only reduction in reserves which occurs arises because non-reserve inputs are substituted for reserves as the cost of reserves increases. Equation (4-9) demonstrates that the reduction in profits is proportionate to the level of expenditures on reserves.

As an alternative to the CONSAD approach, assume that P and Q are allowed to adjust, and that the change in reserves which results is determined by a new supply-demand equilibrium in which the normal profit position of firms is maintained. Assume a demand curve for production with elasticity h.

$$\frac{dQ}{Q} = -h \frac{dP}{P} \tag{4-10}$$

Utilizing Equations (4-2) and (4-10) together with the model of the preceding chapter, one obtains the following result:

$$\frac{dM}{M} = \left(\frac{h \, S_m + s \, S_n}{h \, S_m + b + s \, S_n}\right) b \frac{du}{u} \, . \tag{4-11}$$

This is the correct equation for determining the change in reserves resulting from a change in the depletion allowance when competitive conditions are assumed.

The change indicated by Equation (4-11) will always exceed the change indicated by Equation (4-6), and it will exceed the change implied by Equation (4-8) for parameter values pertinent to the case of petroleum reserves.[b]

This disproves the CONSAD assertion that their procedure overestimates the true reserve impact. The reason for the difference is straightforward. When the supply curve for an input shifts upward due to a tax increase, use of the input will change for three reasons: first, because firms substitute other inputs for the taxed input (whose price is increased); second, because the quantity of the product demanded declines as the increased price of the product results in a movement backward along the demand curve; and third, because the price paid to suppliers of the input (net of the tax) decreases and production of that input declines along its supply curve. The role of these separate influences is apparent in Equation (4-11), which contains the elasticity of product demand, h, the elasticity of input substitution, s, and the elasticity of input supply, b. Of course, the magnitude of the effects depends on the relative importance of the input; that is the reason for the appearance of factor shares in Equation (4-11). The CONSAD analysis identifies only the substitution effect as shown in Equation (4-8).

That this analysis is an accurate appraisal of the CONSAD procedure is verified by the following equation taken from the CONSAD study and written with the notation used in that study. [6]

$$K = \beta \left(\frac{S}{C}\right)^\gamma P^\delta$$

In this equation β, γ, and δ are parameters, S is the price of output, C is the user cost of reserves, P is the quantity of output, and K is reserves. This equation is simply the first order condition for K derived from a CES production function assuming competitive profit maximization. Though it is never mentioned in the CONSAD study, γ is the elasticity of substitution between reserves and other inputs and is the same as parameter s in Equation (4-8). They test several lags for prices in estimating the equation for K but are unable to obtain a value of γ which is significantly different from zero. Since none of their estimates is statistically significant, they choose the largest value obtained in any of their equations and add one standard deviation to this value. This estimate of γ is then treated as the elasticity of demand for reserves. This confirms that the only effect on reserve holdings using their procedure is via substitution of non-reserve inputs for reserves.

[b]Parameter values could be chosen for which the reserve impact indicated by Equation (4-9) would exceed that implied by Equation (4-11). However, it will be shown below that the CONSAD estimate of s is not significantly different from zero. Since all of the parameters in Equation (4-11) are positive, the reserve impact indicated by Equation (4-11) will dominate the change implied by Equation (4-9) when s is near zero.

In principle the idea of treating reserves as an input in production has considerable appeal. As explained in Chapter 2, proved reserves are an estimate of oil which will be recovered from known reservoirs under existing economic and technical conditions. From a conceptual standpoint reserves, so defined, are not an input in the conventional sense. Other inputs can be applied to extract more oil from known reservoirs and, as that is done, proved reserves will be increased. Proved reserves, as a measure of expected production, are an output from that investment, not an input. There is undoubtedly potential for intertemporal substitution of production from a given body of reserves, but that is not considered in the CONSAD study. It is thus not surprising that CONSAD did not obtain an estimate of the elasticity of substitution significantly different from zero.

If one were able to explain econometrically the level of reserves held by the domestic petroleum industry, the major part of specifying an empirical model of supply would be completed. While on the surface the CONSAD model appears to achieve this objective, a more careful look reveals that this model has ignored the most important parameters affecting the level of reserve holdings—the elasticity of supply of reserves and the elasticity of demand for crude petroleum. The model is therefore unsatisfactory for analyzing the elasticity of supply of crude petroleum in the United States, or for assessing the impact on that supply of changes in tax policy.

CONSAD also used two additional methods in attempting to explain reserve holdings. They state, "The second approach utilized a more behaviorally oriented model in which a series of time-dependent relationships between a number of possibly relevant variables (such as gross income, numbers of wells drilled, barrels of reserves discovered, and expenditures for exploration and development) are derived based on a combination of economic theory and empirical data."[7] Their primary effort using this approach was in regressing exploration and development expenditures against a rate of return variable with varying lags. Since they felt that existing estimates of the rate of return in crude oil production were unreliable, they constructed a measure of their own defined as: [8]

$$\frac{\text{Gross Income} - \text{Operating Expenditures}}{\text{Sum of Three Years Expenditures for Exploration and Development}}$$

The trouble with this measure is that it is likely to be highest when the rate of return on new investments is lowest. An oil reservoir typically produces for a period of twenty-five years or more so net income will fall very slowly when exploration and development expenditures are reduced. The CONSAD rate of return measure will be highest when new investments are lowest, but new investments will be low when the marginal rate of return on those investments is low. The CONSAD estimate of the rate of return will therefore move in a direction opposite that of the marginal rate of return perceived by firms in the industry. In comparing a plot of annual expenditures for exploration and development with their calculated rate of return they state: "It can be seen from the data that these two measures do not follow any similar pattern. In fact they appear to follow opposite trends."[9] Not surprisingly, regressions using this variable generally yielded coefficients which were insignificant or of the "wrong" sign. Therefore, CONSAD abandoned this approach as being "infeasible at the present time."[10]

The third approach adopted by CONSAD was to develop a computer model to simulate the behavior of a representative firm in response to a change in the depletion allowance. The representative firm is assumed to have three refineries with *fixed output.* [11] The firm imports the maximum amount allowed by the import quota and produces from its own wells up to the point where marginal production costs equal the market price of crude oil. It then purchases the remainder needed for its refinery inputs. The program is used to compare nine possible ten-year spending programs by the firm for exploration and development. The objective is to choose that spending program which yields the largest value of discounted profits for the firm. Four alternative tax policies are tested for two different types of firms. The firms differ in that one is producing all of its crude refinery inputs while the other is producing 60 percent of its refinery inputs and purchasing the remainder.

The model assumes a fixed level of discoveries per dollar of exploration expenditure.[12] The rate of production from those discoveries is then a somewhat complex function of development expenditures relative to past exploration expenditures. Profitability is the discounted value of the firm's income during the ten-year period plus the discounted value of its reserve holdings at the end of the period. A production restriction of 120 days annually (i.e., one-third of desired production) is imposed on wells producing more than 50 barrels per day.[c] This part of the CONSAD analysis concludes that exploration expenditures are slightly reduced by the elimination of the depletion allowance while development expenditures are reduced by a relatively moderate amount.

This third CONSAD model is also logically inconsistent with competitive profit-maximizing behavior. Suppose initially that the firm is in equilibrium in that its revenue is sufficient to cover exploration, development, and production expenditures and earn a normal profit. Suppose now that price is reduced by a change in the depletion allowance. In the short run, the firm will have reservoirs of varying vintages. If price is sufficient to cover production costs, it will

---

[c]Taken literally, this implies that a well capable of producing 60 barrels per day would be restricted to producing 20 while a well capable of producing 50 would be allowed to produce 50. This does not correspond to the way production restrictions are administered and would render their results meaningless if they followed this procedure in their calculations. Whether this is actually done in the simulation is difficult to tell.

continue to produce from those reservoirs and may incur further development costs. The firm will cease exploration, however, because the present value of its earnings after deducting development and production costs will not be sufficient to cover the cost of making new discoveries while providing a normal rate of return. Since the supply of new discoveries is perfectly elastic (discoveries per dollar of exploration expenditure are constant), the firm will not merely reduce its operations but will in fact go out of business in the long run. Put more simply, the model implicitly assumes that both the demand and the supply curves are perfectly elastic in the long run. If, for some reason, such a market were initially in equilibrium, any exogenous change which shifted the demand curve vertically downward without affecting the supply curve would eventually result in a complete cessation of production in that market.

The CONSAD model failed to reveal this problem for two reasons. First, the model is not really a profit-maximizing model since the time stream of exploration and development expenditures is not chosen to maximize expected profits. Instead, the model chooses the relatively most profitable of nine predetermined alternative expenditure patterns; the idea of a normal rate of profit is not incorporated in the model. The second problem is that the model attempts to assess industry behavior by a partial equilibrium analysis of an individual firm. It should not be surprising that such an approach does not yield results applicable to the entire industry.

THE FISHER MODEL

Fisher's model is the first attempt to econometrically estimate supply equations for any aspect of the U.S. petroleum industry.[13] The influence of his model is clearly evident in all subsequent empirical studies. In the discussion which follows I will identify what appear to me to be shortcomings of Fisher's model. It should be clear, however, that those comments are not intended to deny the importance of Fisher's model as a major development in the analysis of petroleum supply.

Fisher's model consists of three equations intended to explain the following three variables:

1. the success ratio of exploratory drilling; that is, the proportion of exploratory wells which discover oil,
2. the average discovery per well, and
3. the number of exploratory wells drilled.

The product of the above three variables is the volume of crude oil discovered. Fisher's equations are linear in logs and each includes a price variable so the sum of the coefficients on price in the three equations is an estimate of the price elasticity of supply of crude oil. Fisher pools data from the five U.S. PAD

(Petroleum Administration for Defense) districts for the period from 1946 to 1955 giving a total of fifty observations.

In addition to price, all three lagged dependent variables are included in each of Fisher's equations on the grounds that they represent information available to explorers about drilling prospects and also serve to distinguish among the petroleum districts since those districts vary greatly in size. The current price of natural gas and lagged average size of natural gas discoveries are included because natural gas is a joint product with crude oil in exploration. Texas shutdown days are included as a measure of state production restrictions. The amount of geophysical and core drilling crew time are included as a measure of information gathering activity in the equation for wells drilled. Finally, dummy variables are included to distinguish among districts for possible differences not measured by the various lagged dependent variables. Fisher allows for changing costs by deflating price by a cost index and by including average well depth in his equations.

The results indicate a generally high proportion of variance explained. The coefficient of price is highly significant in the wells drilled and average size equation though not in the success ratio equation. The sum of the coefficients of log price is on the order of 0.3 indicating a modest price elasticity. Though his arguments are somewhat involved, Fisher is generally able to rationalize the coefficients for those lagged dependent variables which are statistically significant.

The most troublesome aspect of Fisher's work is that his econometric specification is not derived from an economic model. His heavy reliance on lagged dependent variables is justified by the argument that they represent, in some way, the explorer's knowledge about prospects in advance of drilling. No attempt was made to anticipate the signs of the coefficients prior to estimation. Instead, the results were rationalized after the equations were estimated as Fisher explicitly acknowledges.[14] Also, it is not clear why the quantity discovered is separated into the three components used as dependent variables in the equations; no explanation is given for this procedure.

This formulation of the problem makes it virtually impossible to determine whether the price elasticity is a short-run or long-run elasticity. Ideally, one would like to be able to identify the equilibrium long-run price elasticity in a dynamic model. However, Fisher states: "We do not solve the estimated equation system to determine such an equilibrium numerically, as the dynamics of the process are *hopelessly* obscured by the district-distinguishing roles of the lagged variables." (Emphasis in original.)[15] The "district-distinguishing role" means that the lagged variables are expected to reflect partially differences among regions despite the presence of the dummy variables.

The second difficulty with the model is that the five petroleum districts being analyzed differ enormously in size. The 1950 (midpoint of Fisher's time series) distribution of production and discoveries by district are shown in table 4-1.

Table 4-1.   Distribution of Discoveries by District

PAD District	Discoveries (Percentage of U.S. Total)
District I	0.0
District II	6.83
District III	82.72
District IV	7.88
District V	2.55

Source: Calculated from data reported in *Reserves of Crude Oil, Natural Gas Liquids, and Natural Gas in the United States and Canada and United States Productive Capacity as of December 31, 1972* (Washington: American Petroleum Institute, 1973).

Note: Fisher's data were taken from estimates of the National Petroleum Council. He does not provide a tabulation of the data. The figures in the above table, calculated from American Petroleum Institute data, should be approximately the same as Fisher's.

The supply curves for these regions are clearly very different, and it seems very doubtful that the price elasticity of new discoveries would be the same in all these districts. If they are not, the elasticity for each region should be weighted by that region's share of total discoveries in estimating the aggregate discovery elasticity. This can be demonstrated as follows. Suppose the supply curve in district i is:

$$Q_i = A_i P^{b_i} \tag{4-12}$$

Aggregate supply is:

$$Q = \Sigma Q_i = \Sigma A_i P^{b_i} \tag{4-13}$$

and, after differentiating and rearranging terms, the aggregate supply elasticity is found to be:

$$b = \frac{P}{Q}\frac{dQ}{dP} = \frac{P}{Q}\ \Sigma\ \frac{b_i A_i P^{b_i}}{P} = \Sigma b_i \frac{Q_i}{Q} \tag{4-14}$$

This verifies the assertion that the aggregate elasticity is the weighted average of the regional elasticities where the weights are each region's share of the total.

Suppose that Equation (4-12) is written in logarithmic form and a regression is run with pooled time-series and cross-section data with dummy variables inserted to allow the intercept to vary across regions while the slope is constrained to be the same across regions in the estimation. If the slopes are in fact different across regions, the expected value of the coefficient on log price will be a simple average of the regional elasticities. It will not be an estimate of the aggregate supply elasticity derived in Equation (4-14) above.

This is, of course, a very simple model since price is assumed to be the same in all regions and no other variables are included. This result will still hold if prices differ among regions by a constant amount. This is probably a close approximation to the truth in petroleum production since transport costs are the main source of price differences among regions. If additional variables are included which differ from region to region, the result no longer holds. The model does suggest, however, that the estimated elasticity in a regression with pooled time-series and cross-section data is likely to be an approximate average of the regional elasticities. If the price elasticity is higher in the larger regions—a reasonable presumption in the case of petroleum discoveries—the estimated elasticity will be too low because the larger regions are not given sufficient weight in the regressions. Given the size distribution of U.S. petroleum districts shown in table 4-1, this problem should not be taken lightly.

Finally, a petroleum discovery is a long-lived asset, and the current price of crude oil may be a very imperfect indicator of the value of a new discovery. This suggests that some discounting variable should be included in the econometric model. No such variable is included in Fisher's equations. Also, petroleum deposits are non-renewable assets, and the model should include some variable to reflect the shifting of the supply curve as the inventory of those deposits is depleted by discovery.

While there are shortcomings in Fisher's analysis, it is nonetheless a model which "works"—at least from an econometric standpoint. Before casting such results aside, it behooves the critic to demonstrate that he can do better, and that is the task to which attention will be directed in later chapters of this study. Before taking up this task, it will be useful to consider briefly the modifications which have been made to the Fisher model by others conducting research in this area.

OTHER ECONOMETRIC MODELS

Erikson. Erickson's model of petroleum discoveries[16] is a direct descendant of the model developed by Fisher. Using Fisher's model, Erickson estimates a supply elasticity of 0.9 after correcting errors in the price series used by Fisher.[17] He then extends the time period to 1959 and gets essentially the same results. Erickson's objective is to analyze the choice of drilling opportunities made by the large integrated firms relative to the smaller independent (i.e., non-integrated) operators. For this reason, he extends the model by including variables which measure the relative importance of the two groups of firms across districts. He also estimates the equations with and without production restrictions to measure the losses associated with those restrictions. He does not, however, make any significant changes in Fisher's model.

Erickson and Spann. Erickson and Spann extend Fisher's model by including an equation for average gas discovery size which is similar to that estimated by Fisher for average size of crude oil discoveries.[18] While these authors develop several interesting interpretations of the results in terms of price and cross-price elasticities, the specification of this model is not significantly different from the Fisher model.

Khazzoom. The first model to substantially change the specification used by Fisher was developed by Khazzoom at the Federal Power Commission.[19] The dependent variables in the model are volume of natural gas discovered and extensions and revisions of estimates of amounts discovered in preceding years.[d] The four independent variables in the equation for discoveries are two-period averages of the ceiling price for natural gas, the price of crude oil, the price of natural gas liquids, and the lagged dependent variable. An alternative version of this equation is tested in which all of the independent variables except the lagged discovery variable are squared and included in the equations.

This model is an improvement over prior specifications for several reasons. First, it may be interpreted as a supply equation expressed in dynamic form, and its specification is thus readily understood as a direct application of economic production theory. Second, in contrast to virtually all other models of petroleum discoveries, the long-run equilibrium price elasticities can be derived from these equations. Finally, the model is estimated for several producing regions which have been chosen for their relative homogeneity in terms of geologic characteristics.

Shortcomings of the model are that it contains no discounting variable and no variables to test for exhaustion of the stock of undiscovered reservoirs. Also, the equations were estimated by ordinary least squares without testing for the presence of serial correlation because the author felt that there were no a priori grounds for expecting serial correlation to be present.[20] This is dubious methodology in any case, but particularly in this model which in many respects resembles a model with adaptive price expectations.[21] In such a model, one would expect on a priori grounds that serial correlation would be present. It should be noted that none of the previous models incorporated a test for serial correlation either. In the presence of lagged dependent variables, serial correlation can results in inconsistent estimates [22] so all of these results are somewhat suspect on purely statistical grounds. On the whole, however, the model by Khazzoom is a substantial improvement over prior specifications.

---

[d]Extensions and revisions, reported annually, reflect adjustments to estimates made the preceding year. As information from development and production becomes available in the years after discovery, estimates of the magnitude of the discovery are improved. As indicated in Chapter 2, it takes a period of four to five years before the size of the discovery is accurately established.

MacAvoy and Pindyck. The work by MacAvoy and Pindyck is much more comprehensive than any prior studies in that all phases of the natural gas industry are modeled from exploration through production, transportation, and distribution.[23] Only the portion of the model dealing with exploration will be considered here. The dependent variables in the equations for exploration are number of exploratory wells drilled, average size of new discoveries of associated gas (i.e., gas found with oil) per well, and average size of new discoveries of non-associated gas per well. Independent variables in the wells equation are three dummy variables to distinguish among regions, lagged total revenue from oil and gas production, lagged average total drilling costs, and the sample variance of discovery size in each region. The independent variables in the equation for non-associated natural gas discoveries are the three regional dummy variables, a three-period average of the price of natural gas, a three-period average of the total costs of drilling, and lagged cumulative wells drilled. The equation for associated discoveries differs from the above only in the inclusion of a three-period average of the price of oil in place of the price of gas. The cumulative drilling variable is included to reflect the rate at which potentially productive areas are being depleted. Its inclusion is an improvement over previous specifications.

The justification for the specification of these equations appears to be, for the most part, intuition. The equation for wells drilled seems to be based on the presumption that drilling is primarily determined by cash flow. For the remaining two equations, the authors state that the signs of the variables cannot be determined a priori. In both of these average discovery size equations, lagged total cost has a positive sign. In the discovery equation for non-associated discoveries, the price of gas has a positive sign while in the equation for associated discoveries, the price of oil is negative. The Durbin-Watson statistics in these equations range from 0.36 to 1.06 indicating a high degree of serial correlation. The authors say that they will correct for this in the future.[24]

It would be highly desirable to have a more explicit statement of the type of behavior assumed for firms producing natural gas. A certain degree of skepticism is warranted when the signs on prices and costs in two of three supply equations cannot be predicted on a priori grounds. The rationale for the use of the cash flow approach in determining exploratory drilling needs justification.

Equilibrium price elasticities cannot be readily derived from this model, and there appears to be some question as to the direction of change of discoveries in response to a change in price. In discussing the patterns of signs of the coefficients the authors state: "These effects can occur in this combination because of aggregation of intensive and extensive drilling across districts; their net impact, as prices increase, is *likely* to be increased discoveries." (Emphasis added.)[25]

A comparison of the Erickson and Spann, Khazzoom, and MacAvoy and Pindyck models was recently published by Pindyck.[26] He found a high degree of instability in the coefficient estimates in the Erickson and Spann and Khazzoom models when data for different periods were used. Also, he found substantial differences in the predictions obtained from the three models. These results could, of course, arise from sampling errors inherent in any statistical analysis, but they may also indicate that the models being used do not adequately describe the underlying supply process.

STUDIES OF COST CURVES

Directly constructing cost curves for any aspect of petroleum production is difficult because data are generally not adequate for such a detailed analysis. There have, however, been some noteworthy attempts along these lines. The best known of these is the study of costs of crude oil production in major producing countries outside the United States by Paul Bradley.[27] Bradley performs a careful and thorough analysis of development and production costs in these areas. However, his methodology is most appropriate in areas where finding costs are a small component of overall expenditures and reservoirs are developed as a unit. He specifically rejects application of this approach in studying costs in the United States where production rates are fixed by state regulations.[28]

Henry Steele constructed a short-run cost curve for production in the United States for 1965.[29] While the analysis is limited only to production from existing reservoirs, it demonstrates convincingly that short-run production is extremely insensitive to changes in price. The methodology requires detailed information on costs in the various geographic regions and does not lend itself to application to problems of exploration and development.

SUMMARY

Existing economic models of crude oil and natural gas supply appear to have been developed on a rather ad hoc basis. While general references are made to economic theory, the rationale for the models is not spelled out clearly. Long-run supply elasticities cannot be derived from most of these models, and it is difficult to interpret price elasticities which are estimated in these models. In the chapter which follows, the theoretical underpinnings for the model used in this study are developed in detail.

Chapter Five

Formulation of the Econometric Model

INTRODUCTION

The unifying element for the topics presented in this chapter is the object of providing an economic foundation for the empirical analysis in Chapter 6. The two types of economic agent whose behavior will be analyzed are the oil prospecting firm and the owner of the rights to mineral resources. In the former category are firms who engage in preliminary exploratory activities, leasing of land, and drilling of exploratory wells. The latter type of agent plays a more passive but nonetheless important role of determining whether exploration is to be permitted to proceed on land which he owns.

In the model to be developed, exploration is viewed as a production process in which firms use inputs (exploratory wells and oil-bearing land) to produce discoveries of crude oil and natural gas. This formulation of the problem makes it possible to use economic production theory to develop a structural model of petroleum exploration. The exhaustible resource aspect of the problem is treated explicitly in the development of the supply function for oil-bearing land.

The chapter begins with a review of the characteristics of models of competitive supply of reproducible and exhaustible resources. This serves as an aid in the specification of a supply function for oil-bearing land and will provide a useful background for later interpretation of the results of the econometric analysis. A decision model for the oil prospecting firm is then developed. This model yields a set of derived demand equations for inputs to exploration and a set of supply equations for crude oil and natural gas discoveries. These equations and the supply function for oil-bearing land are then combined, and the equations of the econometric model are derived.

For empirical testing of the model estimates of the value of new discoveries of crude oil and natural gas are required. The procedure for deriving these discovery values is presented, and the effects of federal taxes and state production restrictions on these discovery values are investigated. The product of the chapter is a set of empirically testable equations for the supply of crude oil and natural gas discoveries.

COMPETITIVE SUPPLY WITH
REPRODUCIBLE RESOURCES

The standard competitive supply model assumes that a homogeneous good is produced under conditions of free entry and exit by firms whose objective is to maximize profits. Profits are maximized when firms choose an output level at which marginal revenue equals marginal cost. If cost curves are such that each firm's optimally chosen output level at any price is small relative to demand at that price, the number of firms will be large, and marginal revenue and price will be indistinguishable to the individual firm. Entry will occur until price equals minimum average cost of the least efficient firm, and bidding for scarce factors will assure that the minimum of the average cost curves of other firms will rise to the same level. When adjustment is complete, a point on the industry supply curve is established.

An increase in price will induce all active firms to move upward along their marginal cost curves, and additional firms will be induced to enter until the least efficient firm is again just able to meet all its expenses by operating at the output level where its cost curve reaches a minimum. If all factors are instantaneously variable, the adjustment will be instantaneous, and the new output level is unambiguously another point on the industry supply curve. If price and cost conditions remain unchanged, that same quantity will be produced each period in all subsequent periods. The quantity produced at the higher price will be no less than that produced previously, and it will normally be greater. The amount of the increase depends on the supply curves of factors used in producing the good, and on possible external economies or diseconomies by which changes in aggregate output affect the productive efficiency of individual firms.

Two characteristics of the industry supply curve are of crucial importance. First, the quantity produced depends only on cost conditions in the industry and on the prevailing price. If the firm's production is large enough to influence price, this relationship of production to price breaks down. If the firm is a monopolist, determination of output requires a knowledge not only of cost conditions but of the entire demand schedule as well. If the firm is one of a few large producers, no amount of information about demand and cost conditions will generally be adequate to determine output; one will have to know the personalities and idiosyncracies of the managers of the firms, and it is not clear whether that additional information will be adequate or how it should be used. In short, a supply curve will exist only if all firms take price as a parameter, and one can be sure that they will do so only if their size relative to the market leaves them no alternative. The second characteristic of the industry supply curve is reproducibility; if cost conditions remain unchanged, the supply curve will be the same in every period.

The supply curve indicates what quantity will be produced for any given price at a given instant in time. When all factors are instantaneously variable, there is no ambiguity in the concept. If one or more factors cannot be varied immediately, it is customary to distinguish the short-run supply curve—the price-output schedule realized when some factors are varied instantaneously while others are fixed—from the long-run supply curve—the price-output schedule realized when sufficient time has elapsed to allow all factors to be fully adjusted. A certain sleight of hand is involved in the definition of the long-run schedule because, conceptually, a supply schedule portrays price-output combinations available at a given instant of time; but time must elapse before price-output combinations on the long-run supply curve can be realized. The long-run supply curve is nonetheless a useful construct, and the ambiguity is reduced somewhat if the curve is thought of as depicting the price-quantity combinations realized when any change in price is anticipated in time for all factor adjustments to be made.

The quantity which is actually produced is determined by the point of intersection of the supply curve with the demand curve for the industry product. The market equilibrium point thus established determines the prevailing price and the quantity produced in that and all subsequent periods unless shifts in demand or cost conditions occur.

COMPETITIVE SUPPLY WITH EXHAUSTIBLE RESOURCES

The conventional model of supply outlined above assumes that all factors are either replenishable or are unchanged by the act of production. Machines (and men) may be replaced as they wear out, and land, properly cultivated, is not damaged by the production of crops. While never completely satisfied, the assumption of replenishability or inexhaustibility is an adequate approximation of reality in many industries. In extractive industries such as petroleum, non-replenishability of the resource input is a crucial characteristic of the supply process.

In this section, the theory of supply of exhaustible resources will be reviewed. Since this theory is less familiar than the conventional model, the discussion will be more extensive than the brief review of the conventional model just concluded.

The present value-maximizing firm engaged in the extraction of an exhaustible resource faces a much more formidable task than the firm engaged in production of goods when all factors are replenishable. Costs will be a function of the current rate of production as before. The complexity arises because costs may also be expected to increase as cumulative production increases. The highest quality and most accessible deposits will be exploited first, and the firm will then have to turn to more remote and inferior quality deposits if extraction is to continue. In addition, there may be limits to the rate of extraction in any given period and limits on the total amount of the resource available. The choice of an optimal rate of production requires that these aspects of costs be taken into account, and future prices and interest rates must be predicted as well.

Optimal behavior requires that the firm determine the time path of production throughout the life of a particular deposit which will maximize the present value of net revenues from the deposit. The conditions which must be satisfied for an optimum (analogous to the first order conditions of static analysis) can be specified for qualitatively general models. These will be illustrated below. In addition to specifying the optimality conditions, one would like to deduce the response of the optimum values to changes in the parameters of the supply or demand equations. Unfortunately, optimality conditions in resource depletion models are generally second order differential equations. The analysis of parameter changes requires the comparison of different time paths of production or price. In response to a parameter change (e.g., a shift in the demand curve), output may increase, decrease, or remain unchanged in all periods, or it may increase in some periods and decrease in others. In addition, cumulative production to the point at which exhaustion occurs—the point at which further extraction is unprofitable—may increase, decrease, or remain unchanged. Therefore, results such as those derived from comparative static analysis of the conventional model frequently cannot be derived for resource depletion models.

The first rigorous analysis of exhaustible resources was developed by Hotelling.[1] Generalizations and more extensive analysis of the model were provided by Richard Gordon [2] and Frederick Peterson.[3] Useful diagrammatic treatments are developed by Herfindahl [4] and Scott.[5] In these works, particularly the work of Peterson, considerable progress has been made in deducing the response of production and prices to changes in policy variables, particularly taxes. A good many questions remain unanswered, however. In the discussion which follows, the conditions for optimal exploitation of an exhaustible resource will be reviewed, first for a relatively simple model, then for a general model. After interpreting these results, a discussion of the response of supply to changes in the level, or rate of change, of price will be presented. The discussion concludes with an assessment of the implications for an empirical analysis of petroleum exploration.

The simplest model of resource depletion is one for which current costs are a function of the current rate of extraction but are independent of cumulative production, and the total quantity of the resource available is fixed.[6] The present value of production from the deposit is:

$$PV = \int_{t_0}^{t_1} \pi(S(t), q(t), t) dt \qquad (5\text{-}1)$$

$$= \int_{t_0}^{t_1} [P(S(t), t)q(t) - C(q(t), t)] e^{-rt} dt$$

where

$$\int_{t_0}^{t_1} q(t) \leqslant \underline{Q} \tag{5-2}$$

In equation (5-1), $q(t)$ is the current rate of extraction from the deposit, $\underline{Q}$ is the total quantity available for production from the deposit, $S(t)$ is total current production by the industry from all deposits, and P, C, and r are respectively price, total cost, and the rate of interest; all of which are assumed known in all future periods. As in the conventional case, firms are assumed to produce a sufficiently small amount that they have no control over price. The maximum of (5-1) subject to the constraint imposed by (5-2) will yield an expression for the rate of production $q(t)$ for all t in the interval t_0 to t_1. The calculus of variations provides the appropriate conditions for a maximum.[7]

The first order condition for a maximum of (5-1) subject to (5-2) is:[a]

$$\frac{\partial}{\partial q} [\pi + \lambda q] - \frac{d}{dt} [\frac{\partial}{\partial \dot{q}} (\pi + \lambda q)] = 0 . \tag{5-3}$$

For the problem at hand, $\dot{q}$ does not appear so the second term of equation (5-3) is zero. Since the firm is assumed to produce a sufficiently small amount that it has no control over price, condition (5-3) can be written as follows:

$$(P - MC)e^{-rt} + \lambda = 0$$

or

$$(P - MC) = \lambda e^{rt} \tag{5-4}$$

With increasing marginal costs, net profit on the marginal unit decreases as the current production rate is increased. The firm can produce the additional unit immediately, or it can defer production to a later time possibly earning a higher net profit on the marginal unit at that time; but it will forego the interest income which it could have earned by producing that unit earlier. For present value to be maximum, discounted net profit at any time, t_a, on the marginal unit

---

[a] Additional requirements corresponding to the second order conditions of static analysis must be met as well as continuity and concavity conditions. Since equation (5-3) is a differential equation, boundary conditions at t_0 and t_1 must be specified for an analytic solution. The authors cited previously were unable to derive analytic solutions except for special cases. Since only qualitative results are of interest here, there is no need for a detailed consideration of these problems.

must equal the discounted value of net profit on the marginal unit produced at any other time, t_b. This requirement for the optimal path is more easily seen by rewriting equation (5-4) as follows:

$$(P - MC)_{t_a} = (P - MC)_{t_b} e^{-r(t_b - t_a)} \qquad (5-5)$$

The effect of the resource constraint—often referred to as the "user cost" effect—is to induce the firm to operate at a point where price is greater than marginal production cost. When there is no resource constraint, $\lambda = 0$, and the static optimality condition follows from equation (5-4).

Still another way to write equation (5-4) is:

$$\frac{d}{dt}(P - MC) = r(P - MC) \qquad (5-6)$$

The right-hand side of (5-6) is clearly positive. This implies, from the left-hand side, that net profit must be increasing over time. If price is constant, but the marginal cost curve is upward sloping, the firm can assure that this condition is met by continually reducing its rate of output. By dividing (5-6) by $(P - MC)$, it is seen that net profit on the marginal unit increases at the rate r per period. If price is increasing so rapidly that marginal net profit is rising at a rate greater than r, the firm will produce nothing since capital gains from holding the asset will exceed interest income from producing and investing the proceeds. If net profit is increasing at a rate less than r (e.g., if price is constant and the cost curve is flat), the firm will produce the entire output immediately.

In the more general case costs may be expected to increase as cumulative production increases. Let $Q(t)$ be cumulative production from the deposit, and $q(t) = dQ/dt$ be the rate of production per period. The total cost function is now of the form $C(q(t), Q(t),t)$. The resource constraint (5-2) could be retained, but nothing is lost by assuming an infinite time horizon with increasing costs as the effective resource constraint.[8] The first order condition in this case is:

$$\frac{-\partial C e^{-rt}}{\partial Q} + r(P - MC)e^{-rt} - e^{-rt}\frac{d}{dt}(P - MC) = 0 \, ,$$

or

$$P - MC = \frac{1}{r}\left[\frac{d}{dt}(P - MC) + \frac{\partial C}{\partial Q}\right] . \qquad (5-7)$$

There are two elements of user cost in this formulation. If the firm produces an additional unit of output in the present, it foregoes production of that output in the future, and it increases the cost of all units produced in the future. Both of these factors restrain current production and increase the divergence between price and marginal production cost.

It is clear from the above optimality conditions that production from a given deposit will not be constant over time even if price does not change. Output from the deposit may be expected to decline over time until exhaustion occurs, or until costs rise to the point that further production is unprofitable. Since this will be true for all deposits, it follows that industry output will not be constant from period to period even if price is constant.[b]

What happens if price rises to a new level and remains at that level? In the simple case in which cumulative production does not affect costs, an increase in price shifts production to earlier periods. Suppose the initial production pattern is optimal; it satisfies equation (5-5). If that pattern is retained after the price increase, the left-hand side of equation (5-5) will be greater than the right-hand side. To restore equality, it will be necessary to increase production (and marginal cost) at time t_a, and to reduce production (and marginal cost) at time t_b.[9] In the more general model with optimality condition (5-7), the effect of a price increase is not at all obvious. However, Peterson analyzed a number of cases numerically and concluded that ". . . a price increase unambiguously accelerates extraction . . ."[10] A price increase will also induce firms to exploit deposits which were unprofitable before the increase adding a further increment to production.

It is apparent from this discussion that supply with exhaustible resources differs from the conventional supply case in several important respects. First, at the optimal production rate, price no longer equals marginal production cost.[c] Second, production in any period is dependent not only on current price and cost conditions but on expected future prices, costs, and discount rates. Third, there is no supply curve reproducible or otherwise. A number of supply curves can be derived in any period by postulating a different pattern for prices in the future. Finally, market equilibrium is characterized not by a point in price-quantity space, but by prices and quantities which are functions of time and clear the market for the resource up to the time that exhaustion occurs.

A common misconception is that the price of an exhaustible resource will rise at the rate of interest. While this is true for the special case in which the initial endowment of the resource is fixed and marginal production cost is zero (see equation 5-4), it will not in general be true. If technical progress in extraction is sufficiently rapid, it is possible that the price of the resource may remain

---

[b]It may appear that the above argument is incomplete since new deposits may be brought into use. If new deposits are available in unlimited supply at constant cost, there is no exhaustibility problem. Otherwise, the argument in the text holds.

[c]Marginal cost may be defined to include the opportunity cost of foregoing production of the good in a subsequent period. Price then equals marginal production cost plus marginal opportunity cost.

constant or even decline for a time. One would expect that, at a given point in time, a price increase would call forth an increased supply of the resource unless the increase in price resulted in the expectation of additional price increases in the future. For a given price one would also expect a decline over time in the quantity of the resource produced because of the effect of cumulative extraction in increasing the costs of production, though, as noted above, technical progress could offset this effect.

PETROLEUM DISCOVERIES AND
RESOURCE EXHAUSTIBILITY

Thus far the exhaustion problem has been cast in terms of optimal depletion of a known resource deposit. As such it is directly applicable to depletion of known petroleum reservoirs. Depletion of known reservoirs is not, however, the most important aspect of exhaustion of petroleum resources. If new reservoirs could be found to replace known reservoirs, and costs of finding and exploiting those reservoirs were the same on average as the costs for known reservoirs, the exhaustible resource problem as it relates to petroleum supply would be of minor interest. Such is clearly not the case.

Potentially productive oil-bearing land is an exhaustible resource. The relevance of the theory of exhaustible resources for the present study thus arises from its implications for the form of the supply function for oil-bearing land. The theory of exhaustible resources can be applied to oil exploration by thinking of an oil-bearing region rather than a particular deposit. Perfect certainty cannot be assumed since an essential characteristic of exploration is lack of information about the location and characteristics of deposits. This problem is not particularly troublesome, however, if one is willing to assume that the probability distribution for deposits in the region is known.[11] The suspension of disbelief required for acceptance of this assumption should not be difficult for anyone who is willing to accept the earlier assumption that all future prices, costs, and discount rates are known.

The rate of exhaustion of potentially productive oil-bearing land is not solely determined by the oil-prospecting firm. It will be argued in the following section of this chapter that, from a theoretical standpoint, the oil prospecting firm is conceptually no different than firms producing other goods and services. The prospecting firm buys or rents inputs (exploratory wells and oil-bearing land) and produces outputs in the form of information about the location of crude oil and natural gas deposits. Petroleum exploration involves a greater degree of uncertainty than production of other types of goods, but that is a difference of degree not of kind.

It is rather the resource owner, the owner of the mineral rights to land, who determines the rate at which exhaustion occurs by his decision either to permit exploration to proceed or to withhold the land from exploration. Though the landowner will be uncertain about the presence or absence of petroleum deposits on his property, his leasing decisions will, in principle, be guided by the same forces which determine the rate of exhaustion of known resource deposits. If he expects that higher land rentals or more generous royalties will be forthcoming in the future, he may choose to withhold the land from exploration. In addition, there may be opportunity costs in the form of foregone income from agricultural production if exploratory drilling is permitted.

In the aggregate the supply function for oil-bearing land will thus have the properties identified in the discussion of exhaustible resource models. For the econometric analysis, the form of the aggregate supply function will be taken to be as follows:

$$L = R^{\beta_1} e^{\gamma_2 W} \tag{5-8}$$

where R is the unit rent on oil-bearing land of a given quality, and W is cumulative past exploratory effort measured by the number of exploratory wells drilled. At a given point in time an increase in R will lead to an increase in the amount of land supplied because the land is bid away from other uses, and because some landowners who were withholding land from exploration in the expectation of an increase in rents will be induced to make the land available. Thus β_1 should be positive. As exploratory drilling takes place, the amount of land thought to be potentially productive of petroleum is reduced either because reservoirs are discovered and exploited or because unsuccessful wells indicate that an area previously thought to contain petroleum is, in fact, unproductive. Therefore, γ_2 should be negative. It should be emphasized that the equation specified above is intended to reflect the *aggregate* supply of oil-bearing land.

The demand for oil-bearing land arises from its use as an input to exploration. In the sections which follow, the equations determining the demand for oil-bearing land and exploratory drilling are derived along with the equations determining the supply of crude oil and natural gas discoveries.

A DECISION MODEL FOR THE
OIL-PROSPECTING FIRM

The analysis of the decisions of the firm exploring for oil and natural gas will be facilitated by assuming that there is a well functioning market for oil and natural gas discoveries. When a discovery is made, the firm may exploit the deposit itself, or it may sell the rights to the discovery to another firm. Let P_o and P_g be the after tax price per unit of oil and gas in the ground. These are not wellhead prices, but rather the unit values of oil and natural gas in the reservoir prior to the drilling of producing wells. The major inputs in the exploration process are exploratory wells and oil-bearing land. Let C be the after tax cost per unit (per foot) of drilling and R be the after tax unit rental for oil-bearing land.

The objective function of the exploration firm is then:

$$\pi = P_o \, Y + P_g \, G - CF - RL \tag{5-9}$$

where Y and G are quantities of oil and gas discovered and F and L are respectively footage drilled and amount of oil-bearing land used. It will be assumed that the technology of oil exploration can be represented by a production function as follows:

$$H(Y,G,F,L,W) = 0 \tag{5-10}$$

W is a trend variable explained below. The role of the prospecting firm is thus conceptually the same as the role of firms producing goods and services in other industries. The firm chooses the input and output combinations which maximize (5-9) subject to the constraint imposed by (5-10). Industry input and output levels are then the sums of the input and output levels chosen by all firms. It is these industry aggregates which are the subject of empirical analysis in Chapter 6.

Ideally, the decision model in equations (5-9) and (5-10) should involve choice of desired output and input levels by maximization of expected profits. In principle, this is quite straightforward. Stochastic elements can be introduced in equation (5-10) to represent the prospector's evaluation of the probability distribution of discovery prospects. Likewise, random elements could be introduced in (5-9) to represent the prospector's subjective evaluation of the distribution of expected prices and costs.

This procedure is feasible in practice as well if the function in (5-10) is not too complex, and the random elements (with judiciously chosen probability distributions) are introduced in equations (5-9) and (5-10) in a "manageable" fashion; manageable meaning that an analytic expression for the mathematical expectation of profits can be derived.[12] The form chosen below for H is not mathematically simple, and the random elements would have to be introduced in a highly artificial fashion to develop a profit function for which an analytic expression for the expected value could be obtained. Rather than attempt such a procedure, the equations for optimal input and output levels are derived from a deterministic model of the form outlined in equations (5-9) and (5-10). In Chapter 6 various procedures for introducing random elements in the equations derived from this model are presented and tested.

THE PRODUCTION FUNCTION APPROACH

The objective of this study is the empirical estimation of the parameters of supply functions for crude oil and natural gas. This can be accomplished either by estimating the parameters of a production function and the associated first order conditions, or by estimating the parameters of appropriately chosen supply equations. A question arises as to which of these approaches is appropriate.

The answer is that the procedures are theoretically equivalent. Functional forms chosen for empirical analysis are necessarily arbitrary; and, whether supply or production functions (or profit or cost functions), they are chosen because it is believed that they are sufficiently general to adequately represent the characteristics of the underlying production technology for a given problem. Restrictions on the underlying technology imply restrictions on all of the above types of functions, and restrictions on any one type of function imply restrictions on the others.[13]

The particular approach which is chosen empirically is thus determined by the available data, the objectives of the research, and the preferences of the researcher. In the present case, the parameters can be obtained either by estimating the parameters of the production function and the related first order conditions, or by estimating the parameters of the implied supply equations. Though the deterministic models are equivalent, the error structures chosen in the two cases will not necessarily be the same. Discussion of this problem is deferred to Chapter 6.

Choice of Functional Form

In choosing a functional form the competing demands which must be reconciled are the desires for computational simplicity, parsimony in the use of parameters, and generality of the function as a representation of the underlying technology. The most general functional specifications available are those which are second order approximations to any production function. Functions in this category are the "generalized linear transformation function"[14] and the "trans-log production frontier."[15] The former includes the variables, the square roots of all the variables, and the cross products of the square roots of all the variables. The trans-log includes the logs of all the variables plus the squares and cross products of the logs. The number of parameters to be estimated is the same in both cases, and both can be estimated using linear simultaneous equation methods. The functions contain a large number of parameters (fourteen for a model with two inputs and two outputs) which must be reduced to a smaller number by the imposition of constraints if statistically significant parameter estimates are to be obtained. In practice this is generally done by testing various restrictions on the parameters and accepting those which do not lead to a significant reduction in explained variance. This is a theoretically appealing procedure, but it should be noted that the generality of these specifications is somewhat illusory since they are local, not global, approximations to any function. In the present case the procedure of testing constraints on general functions is computationally costly since expected prices based on a geometric weighting of past prices are to be used in the analysis. With this treatment of expected prices the econometric equations are no longer linear in the parameters, and tests of various parameter restrictions could be conducted only at considerable computational cost.

The alternative approach, adopted for this study, is to begin with a simpler function and to test the effects of relaxing restrictions on that function. The simplest production function is the Cobb-Douglas, but it is unacceptable for a problem with multiple outputs because it implies a unitary elasticity of substitution among all inputs and outputs. Specifically, this yields a transformation frontier between the outputs which is convex to the origin. A transformation curve with that property would result in complete specialization in one output—it would not be a joint output problem.[16]

The simplest functional form which is adequate to represent a problem with joint outputs is the "Constant Elasticity of Transformation" (CET) function derived by Powell and Gruen.[17] This function, written in a form suitable for the present study, is:

$$[a\,Y^d + (1-a)G^d]^{\frac{1}{d}} = A_1 F^m L^n e^{\gamma_1 W} \qquad (5\text{-}11)$$

where Y and G are discoveries of oil and natural gas respectively, F and L are inputs of drilling and oil-bearing land respectively, and W is a trend variable included to measure technical progress. This trend variable will also reflect the use of geological and geophysical methods since it was argued in Chapter 2 that these methods enhance the productivity of exploratory wells in much the same way as advances in exploration technology increase drilling productivity. It was also noted in Chapter 2 that changes in exploration technology have been of an evolutionary nature during the period in question so the trend variable should adequately reflect the effects of advances in technology.

Cumulative wells drilled will be used as the trend variable on the grounds that technical knowledge and technical improvements in exploratory methods arise primarily as a result of innovations made in the actual course of exploration. Elsewhere this has been referred to as the hypothesis that technical progress is a result of "learning by doing."[18] An alternative view is that technical progress arises primarily by the growth of knowledge through time and largely independent of the level of activity in any particular industry. As a practical matter, time and cumulative wells are so highly correlated that one could not hope to distinguish between these hypotheses empirically. The truth is probably somewhere in between anyway, and the cumulative wells variable can be expected to pick up both effects.

The input side of equation (5-11) is mathematically identical to the Constant Elasticity of Substitution (CES) production function.[19] However, for the CET, the second order conditions require that d be greater than one whereas, for the CES function, the second order conditions require that d be less than one.

The value of d determines whether the isoquants are concave or convex to the origin. This function is restrictive not only because it implies a constant elasticity of transformation between the outputs, but also because it has homothetic isoquants. This implies equal supply elasticities for Y and G—an undesirable a priori restriction on the supply equations. The following more general function based on the work of Hanoch [20] does not impose these restrictions.

$$(a) Y^{d_1} F^{d_1 m_1} L^{d_1 n_1} e^{-d_1 g_1 W} + (1-a) G^{d_2} F^{d_2 m_2} L^{d_2 n_2} e^{-d_2 g_2 W} = B \qquad (5-12)$$

This function reduces to the form of equation (5-11) when $d_1 = d_2$, $m_1 = m_2$, and $g_1 = g_2$. This function is less convenient than the preceding one because the implied supply and derived demand equations cannot be written explicitly. The procedure used in Chapter 6 is to first estimate the parameters in equation (5-11) and then to test whether a statistically significant improvement is obtained by use of equation (5-12).

While these functions afford a considerable degree of generality with a moderate number of parameters, computational simplicity is sacrificed because the functions are non-linear. The first order conditions for both (5-11) and (5-12) are linear in logs, however, so preliminary tests can be made using standard linear methods.

DERIVATION OF THE STRUCTURAL EQUATIONS

Each oil-prospecting firm is assumed to have a production function, represented by H in equation (5-11). That function may or may not be the same for all firms. Each firm determines its input and output levels by maximizing profits subject to its own production function taking prices and factor costs as parameters. Aggregation of those micro-relationships yields the industry input and output equations. Maximizing equation (5-9) subject to the production function (5-11) or (5-12) treating prices as parameters also yields a set of input and output equations. It is assumed in the empirical analysis that the equations thus obtained are an accurate representation of the equations which would be obtained by aggregating the micro-relationships.

The first order conditions obtained by maximizing (5-9) subject to the constraint (5-11) are the following:

$$(a) \quad P_O - \lambda [(a)Y^d + (1-a)G^d]^{\frac{1-d}{d}} (a)Y^{d-1} = 0$$

$$(b) \quad P_g - \lambda [(a)Y^d + (1-a)G^d]^{\frac{1-d}{d}} (1-a)G^{d-1} = 0 \qquad (5-13)$$

(c) $-C + \lambda A_1 m F^{m-1} L^n e^{\gamma_1 W} = 0$

(d) $-R + \lambda A_1 n F^m L^{n-1} e^{\gamma_1 W} = 0$

where λ is the Lagrangian multiplier. Equations (5-8), (5-11), and (5-13) are the six equations in the six endogenous variables Y, G, F, L, R, and λ. By taking ratios of the equations in (5-13), the Lagrangian multiplier can be eliminated. Neither the price, R, nor the quantity, L, of oil-bearing land is observable. These variables can be algebraically eliminated by using equations (5-8), (5-11), and (5-13). The resulting equations are as follows:

(a) $[(a)Y^d + (1-a)G^d]^{\frac{1}{d}} = AF^{m+\beta}\left(\dfrac{C}{m}\right)^\beta e^{g W}$

(b) $\dfrac{Y}{G} = \left(\dfrac{1-a}{a}\right)^{\frac{1}{d-1}}\left(\dfrac{P_o}{P_g}\right)^{\frac{1}{d-1}}$ (5-14)

(c) $\dfrac{CF}{P_o Y} = m + m\left(\dfrac{1-a}{a}\right)\left(\dfrac{G}{Y}\right)^d$

where

$g = \gamma_1 + \dfrac{\gamma_2 n}{(\beta_1 + 1)}$, $\beta = \dfrac{n\beta_1}{(\beta_1 + 1)}$, and $A = A_1 n^\beta$.

Parameter g thus measures the net effect of technical progress and exhaustion.

By eliminating the endogenous variables R and L, a set of "partially reduced form equations" is obtained. As the following statement indicates, this type of model characterizes most econometric analysis:

Equations obtained by simultaneously eliminating one or more equations and one or more endogenous variables have been called partially reduced form equations in various discussions. In a certain fundamental sense, all equations we are likely to deal with may be regarded as partially reduced form equations. It is always possible to imagine a more fundamental explanation of the phenomena that we observe, involving more equations and more endogenous variables. If the model we use is a reasonable one, it should, in principle, be possible to derive it, either exactly or approximately, from the more fundamental model by successive elimination of variables.[21]

This is precisely the logic guiding the development of the above set of equations.

It was emphasized at the outset of the preceding derivation that P_o and P_g are not wellhead prices but rather the unit value of new discoveries in the ground. To state the equations in an empirically testable form, it is necessary to develop a procedure for estimating these prices. This problem is taken up in the following sections, and the effects of federal taxes and state production restrictions on these discovery prices are investigated.

THE PRICE OF A DISCOVERY

Though there is an active market for petroleum discoveries, the results of discovery transactions are not systematically reported. By adopting some simplifying assumptions, however, an estimate of the unit value of a discovery can be made. A firm making a decision about the amount of leasing and exploratory drilling to undertake must make such an estimate as well. The following derivation should be viewed from the perspective of a firm which does not know either the prices and discount rates that will prevail in the future or the precise characteristics of reservoirs which it may discover. The simplifying assumptions adopted below are probably an accurate reflection of the approximations made in practice when currently available information is used to estimate the value of deposits which may be discovered. The results derived below will be applied to obtain separate price series for crude oil and natural gas discoveries. Except where indicated, the same assumptions will be applied in calculating both price series.[d]

The net present value of a discovery is:

$$NPV = \int_0^\infty P_w(t)q(t)e^{-rt}dt - \int_0^\infty C(q(t),t)e^{-rt}dt$$

$$-T[\int_0^\infty P_w(t)q(t)e^{-rt}dt - \int_0^\infty C(q(t),t)e^{-rt}dt$$

$$-d\int_0^\infty P_w(t)q(t)e^{-rt}dt] ,$$

or

[d]Certain interdependencies between oil and gas production are necessarily ignored in this derivation as are problems of aggregation. Each oil or gas reservoir will have a separate production curve depending on natural drive characteristics, local drilling costs, gas-oil ratios, etc. as explained in Chapter 2. In the calculations below, all oil reservoirs are assumed to have similar production curves corresponding to the observed aggregate as are all gas reservoirs. Again, some such simplifications must also be made by a firm attempting to estimate the value of reservoirs which may be discovered.

$$NPV = (1-T+Td)\int_0^\infty P_w(t)q(t)e^{-rt} - (1-T)\int_0^\infty C(q(t),t)e^{-rt}dt. \qquad (5\text{-}15)$$

In this expression, $q(t)$ is production of oil or gas from the reservoir t years after discovery, $P_w(t)$ is the wellhead price (of oil or gas), r is the rate of discount, C is the cost of development and production, and T and d are respectively the corporate profits tax rate and the depletion allowance rate. Exploration costs are not included here since the objective is to obtain an estimate of the value of a reservoir after it has been found. This formulation assumes that the firm's tax deductions do not exceed its taxable income and that the depletion allowance for the reservoir does not exceed 50 percent of net income on the property at any time during the life of the reservoir. The validity of these assumptions will be discussed later.

The evidence reviewed in Chapter 2 supports the assumption that the rate of production decline from a reservoir is approximately exponential.

$$q(t) = q_0 e^{-Dt} \qquad (5\text{-}16)$$

The decline rate, D, is the fraction by which production declines from one period to the next. Cumulative production to the point of exhaustion of the reservoir is the integral over time of equation (5-16).

$$Q_c = \int_0^\infty q_0 e^{-dt}\, dt = q_0/D \qquad (5\text{-}17)$$

If cumulative production is proportional to the volume of oil or natural gas discovered, then

$$Q_c = kQ \qquad (5\text{-}18)$$

and, from (5-17) and (5-18):

$$q_0 = DkQ. \qquad (5\text{-}19)$$

In equation (5-19), Q is either original oil-in-place discovered or recoverable natural gas. For oil, k is approximately 0.30, and for natural gas, approximately 1.0. Proved reserves are an estimate of oil or gas remaining to be produced at time t.

$$RES(t) = \int_t^\infty q_0 e^{-Ds}ds = \frac{q_0}{D}\, e^{-Dt} \qquad (5\text{-}20)$$

From equations (5-16) and (5-20) it follows that

$$D = \frac{q(t)}{RES(t)} \; ; \qquad (5\text{-}21)$$

the decline rate equals the production-reserves ratio.

If the price and the discount rate are set at constant expected values, the first term in equation (5-15) can be written as

$$GPV = (1-T+Td)\,(P_w DkQ)\,/\,(D+r). \tag{5-22}$$

In the above expression, the term $\dfrac{D}{(D+r)}$ is the present value of an annuity which pays D in the initial period and declines at the rate of D per period. The undiscounted value of such an annuity is one dollar. A reservoir which declines exponentially at rate D is an annuity which pays $D(1-T+Td)\,P_w kQ$ in the initial period and declines by $100D$ percent per period in each subsequent period. Equation (5-22) is thus the present value of gross revenues from a petroleum discovery. Gross revenue per unit discovered is then $\dfrac{GPV}{Q} = (1-T+Td)$ $(P_w Dk)\,/\,(D+r)$.

The terms in these expressions represent the oil prospector's estimate of values of those variables which will prevail in the future, but all have an empirically measurable counterpart at any point in time. More will be said about the determination of expected prices in the following chapter.

Unfortunately, the cost integral in equation (5-15) is not readily converted into an observable quantity. However, some evidence is available concerning the proportion of total production costs attributable to exploration. Table 5-1 indicates the proportion of costs in each category for the years for which data are available. Over the long run in a competitive industry, average total costs should equal average revenues with due allowance for random elements. Thus, on average, the proportion of revenues from oil production which are attributable to exploration should equal the proportion of total extraction costs which are attributable to exploration.[e] From the data in table 5-1 it appears that the proportion spent for exploration has been approximately constant, or perhaps slightly increasing, over the 1948-68 period.

Since the proportion of revenues attributable to exploration is not observable, it will be left as a parameter for empirical estimation, and the possibility of a trend in the share accorded to exploration will also be tested. If b_1 is the proportion attributable to exploration in the initial period, and h is the trend in that proportion, equation (5-15) can be written as follows using equation (5-22).

$$NPV = b_1 e^{ht}\,(1-T+Td)\,\frac{(P_w DkQ)}{(D+r)} \tag{5-23}$$

[e]This will be true only if no economies are realized in conducting exploration, development, and production together. Since these operations are necessarily sequential, the only potential technological savings from conducting them together would be in overhead expenses.

Table 5-1.   Proportion of Production Costs Attributable to Each Stage
of Extraction

	Exploration	*Development*	*Production*
1948	31	30	39
1953	32	30	38
1955	28	34	38
1956	29	33	38
1959	32	32	36
1960	32	34	34
1961	34	31	35
1962	32	34	34
1963	36	30	34
1964	36	32	34
1965	36	31	33
1966	33	32	36
1967	34	31	35
1968	38	28	33

Source: Computed from *Joint Association Survey of Industry Drilling Costs* (Washington: American Petroleum Institute). The survey was conducted intermittently prior to 1959. Further detail is provided in Appendix B.

The unit price of a discovery is then: $P = \dfrac{NPV}{Q}$.

A preliminary estimate of b_1 can be obtained using the information in table 5-1. The value of b_1 cannot be taken directly from the table because, for a particular reservoir, development and production costs are incurred subsequent to exploration. Development and production costs must be discounted relative to exploration costs, and the proportion of total costs accorded exploration is higher than the data in table 5-1 would indicate. When this discounting is taken into account, the average share attributable to exploration is 0.41 for the entire period. The details of this estimate of b_1 are reported in Appendix B. This value will be useful in preliminary testing of the model.

In computing the unit price of a natural gas discovery one exception is made to the above procedure. When it is produced at the wellhead, much natural gas is "wet"; liquid hydrocarbons are dissolved in the gas. The unit price of gas must therefore be adjusted for the value of the liquids which are recovered. This is non-trivial since, during the early years of the period being analyzed, the value of marketed natural gas liquids equaled or exceeded the value of marketed natural gas.

Fortunately, the proportion of natural gas liquids to natural gas produced has remained approximately constant during the period being analyzed. Utilizing the assumption of fixed coefficients, the price per unit of natural gas discovered is equal to the weighted average of the price of natural gas and the price of natural gas liquids. The weight assigned to each is determined by the proportions in which the two outputs are produced. The details of the calculation are reported in Appendix B.

It is not immediately apparent, but the price series provided by equation (5–23) does incorporate the effects of state production restrictions. This may be illustrated by recourse to a discrete formulation of equation (5–16). In the absence of administrative restrictions, desired production at time t will be a fixed proportion, D', of oil remaining in the reservoir at the beginning of that period if production decline is exponential. At time zero, no oil has been removed so production is:

$$q_0 = D'kQ \tag{5-24a}$$

Where D' is the desired decline rate. In period one, the remaining oil is $kQ-q_0$ so period one production is:

$$\dot{q}_1 = D'kQ(1-D'). \tag{5-24b}$$

By continuing this procedure, it is found that production in period t is:

$$q_t = D'kQ(1-D')^t = q_0(1-D')^t. \tag{5-24c}$$

which is the discrete counterpart of equation (5–16).

Now, suppose that each period a regulatory authority restricts output to a fraction, x, of desired production. Then production in period zero will be:

$$q_0 = xD'kQ. \tag{5-25a}$$

In period one, the remaining oil in the reservoir is $kQ-xD'kQ$ so period one production is:

$$q_1 = xD'kQ(1-xD'). \tag{5-25b}$$

Extending this sequence of steps, it is found that period t production is:

$$q_t = xD'kQ(1-xD')^t = q_0(1-xD')^t. \tag{5-25c}$$

in the presence of production restrictions. Reserves in this case are:

$$RES_t = \sum_{s=t}^{\infty} q_s = q_0 \sum_{s=t}^{\infty} (1-xD')^s = \frac{q_0}{xD'} (1 - xD')^t \tag{5-26}$$

Dividing equation (5–25c) by equation (5–26) it is found that:

$$xD' = \frac{q_t}{RES_t} = D. \qquad\qquad (5\text{-}27)$$

The observed decline rate, *D,* thus incorporates the effects of production restrictions. As the allowable production rate, *x,* is reduced, the observed decline rate is reduced. This in turn reduces the present value of a petroleum discovery as can be seen from equation (5-23). If the discount rate were zero in equation (5-23), the decline rate would not matter since a dollar in the future is worth as much as a dollar in the present. However, for a positive discount rate, any reduction in the decline rate reduces the value of the discovery because the receipt of revenues from the discovery is postponed. Equation (5-23) thus provides a suitable measure of the unit price of a petroleum discovery in the presence or absence of production restrictions.

It should be emphasized once more that the price series in equation (5-23) is intended to reflect the expectations of the firm engaged in oil exploration about the per barrel income from a possible discovery. Several assumptions were made either explicitly or implicitly in the derivation of equation (5-23). The validity of these assumptions will now be discussed.

1. It has been assumed that the depletion allowance on a reservoir does not exceed 50 percent of net income from the reservoir in any year. While little empirical evidence is available on this question, data cited in Chapter 2 indicated that the average rate of depletion actually taken by oil producing firms was approximately 25 percent during the early 1950s when the statutory rate was 27.5 percent.[22] Thus a small proportion of existing reservoirs—those with high cost production—were encountering the limitation on net income. Only reservoirs in the late stages of production when costs are high relative to price will encounter the 50 percent income limit. At the time of discovery, the loss of such tax deductions will be negligible in present value.

2. It is assumed that firms engaged in oil exploration have taxable income which exceeds their tax deductions. All tax deductions available to oil producing firms except the depletion allowance may be applied against income from any source. Firms which persistently have tax deductions which exceed their taxable income have an incentive to merge with firms who have income to which the tax deductions could be applied. In part, the trend toward increasing vertical integration in the industry is probably attributable to the desire to fully take advantage of the tax deductions in oil exploration and production. While this is an area where systematic study is needed, there are probably very few producing firms which are not in a position to fully take advantage of their tax deductions.

3. Current and past aggregate values of price, interest rate, decline rate, and recovery rate provide adequate information about the values of those variables expected by firms to prevail during the production period for any reservoir which may be discovered. It seems fair to assume that firms form their expectations of the future on the basis of present and past information, since any individual having certain knowledge of the future could find a more

remunerative and less demanding occupation than petroleum exploration. Whether published aggregates can be used to provide an adequate guide to those expectations is far less certain. In the absence of better information, the decision must rest on whether reasonable empirical estimates are obtained with the available data.

4. State regulations can be adequately represented as restricting production to a fixed fraction of desired output. This is a simplification of the actual regulatory process. As explained in Chapter 2, actual production restrictions are based on depth and spacing of wells in a reservoir and are not directly tied to the productive capacity of the reservoir at a given point in time. The regulations are thus likely to be more onerous during the early years of production from a reservoir than the above assumption would suggest. However, most state regulatory agencies have established so-called discovery allowables which permit a relatively high rate of production during the first year to eighteen months after discovery. Therefore, when the entire production stream from the reservoir is considered, the assumption that allowable production is a fixed fraction of desired production is probably a relatively close approximation of the actual effect of production restrictions on the present value of a discovery.

EQUILIBRIUM IN THE MARKET FOR DISCOVERIES

There is no question that one objective of state regulations has been to maintain price above the level that would prevail in the absence of state intervention. Equality of supply and demand at the chosen price at the production stage has been achieved by direct intervention; the state has determined allowable production. While there was no direct intervention at the exploration stage, changes in price and the allowable production rate influenced exploration incentives. This is most easily seen by substituting equation (5-27) into equation (5-23).

$$NPV = b_1 e^{ht} (1 - T + Td)(P_w x D' kQ) / (xD' + r). \tag{5-28}$$

When there is no intervention at the production stage, $x=1$. Now suppose price is raised administratively to a price above the equilibrium level, say P'_w. In practice, this increase in price would have been effected by setting x to a value less than one. The increase in price would tend to stimulate exploration while the reduction in the allowable production rate would tend to discourage exploratory activity as explained previously. The net effect initially is likely to be an increase in exploration and a consequent increase in desired production. To maintain P'_w, x would have to be reduced still further, and this in turn would reduce the value of a discovery. This process would continue until new discoveries were just sufficient to provide the allowed production at the chosen price, P'_w.

The result is that, when price is fixed, the allowable production rate, x, takes the role of price in adjusting the level of exploratory effort. The consequences of this intervention are a particularly poignant example of how the market "takes its revenge upon those who would defy it," and the result is in accord with Baumol's observation that "the punishment often fits the crime."[23] In the middle fifties and early sixties, states which adopted the most onerous output restrictions (particularly Texas and Louisiana) found that exploratory effort in their states was declining more rapidly than in states which had less stringent restrictions. In an effort to compensate, these states instituted "discovery allowables" which provided exemption from the production restrictions for the first year or eighteen months after discovery. Those efforts were necessarily self-defeating, however, because success in stimulating exploration meant an increase in productive capacity which in turn required still more onerous output restrictions in order for the artificially high price to be maintained.

THE ROLE OF TAXES

The effect of a change in the depletion allowance rate or the corporate tax rate is readily apparent from equation (5-28). An increase in the corporate tax rate reduces the value of a discovery, other things constant. In contrast, an increase in the depletion allowance rate increases the value of a discovery as long as the firm has taxable income and the 50 percent income limit is not reached. Of course, other things will not be constant in a general equilibrium context. A reduction in the tax rate would induce greater exploration because of the increased profitability of a discovery. The resulting increase in production would drive prices down, or, if price were maintained by state intervention, the increase in desired production would force a lowering of x. These general equilibrium effects are registered in the observed price and production data.

DRILLING COSTS

The IPAA index of drilling cost per foot will be used as a measure of the unit cost of drilling. Since it is an index, it must be converted to cost per foot by multiplying it by a constant. Thus:

$$C = b_2 C_I \tag{5-29}$$

where C_I is the IPAA cost index with a base of 1.00 in 1947. Data reported in Appendix B indicate that b_2 is approximately 8.40, which implies that drilling cost per foot was $8.40 in 1947. Information in that appendix indicates that the IPAA index probably overstates the rate of increase of drilling costs. This is because the IPAA index is partially based on an index of the costs of various inputs. It does not fully reflect cost savings which arise due to more efficient use of those inputs.

By referring to the estimating equations derived earlier, it will be seen that the unit cost of drilling, C, appears in a ratio to either the price of oil or the price of natural gas. Therefore, the parameters b_1 and b_2 introduced above can be summarized by a single parameter $b = .41\dfrac{b_2}{b_1}$ which will reflect errors in estimating the unit cost of drilling in the base period as well as errors in estimating the share of revenues attributable to exploration.[f] Likewise, the parameter, h, in equation (5-23) will reflect any trends in actual drilling costs relative to the IPAA index as well as trends in the share of revenues attributable to exploration. Both of these trends have the same sign so h should be positive.

SOME METHODOLOGICAL NOTES

In equation (5-14), P_o P_g, and C have been interpreted as expected discovery prices and expected costs. As a practical matter, the alternatives for measuring expected prices and costs are either to take current values as an adequate measure of expected values or to use some weighting of past prices as a measure of expectations. The latter procedure is generally preferable. Results for both procedures will be reported in Chapter 6.

Thus far the aggregation problem has been dealt with by assuming it away. This is a common, if unsatisfying, procedure. In the present case, the most desirable type of data, data for individual firms, are not available. As explained in Chapter 2, data are reported on a regional basis only; firms are given complete anonymity in the published tabulations. Thus, problems of aggregation are present even if regional data are used. Data for these regions could be pooled to increase the sample size. However, as was argued in connection with the Fisher model in Chapter 4, this type of pooling differs in important respects from pooling of data for individual firms. All firms in an industry may be assumed to have access to the same production technology, so pooling of data for firms is a highly desirable procedure in most cases. Where data are aggregated to the level of producing regions with arbitrarily drawn boundaries, supply conditions will differ from region to region because endowments of oil-bearing land will differ from region to region. Pooling data under such circumstances is an indefensible procedure.

The alternative to pooling is to estimate separate supply equations for each producing region. If different regions have differing endowments of oil and natural gas, firms may respond to changes in the relative prices of those outputs by moving from one producing region to another. This effect will not be measured if separate equations are estimated for each region. Despite this drawback, estimation of separate equations for each region is likely to yield useful information. Before embarking on a project of that magnitude, however,

---

[f]If the initial estimate of $b_1 = 0.41$ cited earlier is correct, b should be approximately equal to b_2.

it seems best to first test the methodology using aggregate data as is done in Chapter 6.

Before turning to the empirical results, one final issue warrants comment. The production functions in equations (5-11) and (5-12) are representations of technology. No behavioral assumptions are required in the specification of those functions. In the derivation of equations (5-14) the assumption of competitive behavior was utilized. Three important advantages are gained from the use of this assumption. First, and most important, behavioral assumptions are imperative if supply predictions are to be made. Perfect knowledge of the form and parameters of the production function and input supply equations is not sufficient for a determination of production. If the assumption of competitive behavior is adopted, current and expected future prices are the requisite additional information. As emphasized at the outset of this chapter, it is not clear what additional information is required for a determination of production under oligopoly conditions.

Of course, convenience is no justification for the use of assumptions which are patently wrong. In the present case, the assumption of competitive behavior seems more tenable than the alternatives afforded by oligopoly theory. If data for individual firms were available, the assumption of competitive behavior could be tested.[24] In the absence of such information, behavioral assumptions must be made on the basis of evidence such as that presented in Chapter 2. In several of the studies reviewed in Chapter 4, no explicit behavioral assumptions were made. It might be thought that the results of those studies are superior to results obtained when the assumption of competitive behavior is used explicitly. Problems do not, however, go away by virtue of being ignored. Behavioral assumptions are necessary as a guide both in specification of equations and in interpreting the results. The rather peculiar specification of several of the models reviewed in Chapter 4 may be attributable to the failure to clearly formulate a behavioral model.

The second advantage of the assumption of competitive behavior is that additional equations are obtained which provide a means of obtaining more precise parameter estimates. Thus, equations (5-14b) and (5-14c) are added to the production function, equation (5-14a). The final advantage of adopting a specific behavioral assumption is that it provides a means for dealing with simultaneity problems in econometric estimation. This will be taken up in the presentation of the empirical analysis in Chapter 6 to which attention is now turned.

Chapter Six

Empirical Results and Conclusions

INTRODUCTION

Simultaneity in the determination of input and output levels characterizes most production decision models. Much of the literature on production function estimation is concerned with determining the effect on parameter estimates of ignoring this simultaneity,[1] in finding circumstances for which simultaneity will not invalidate single equation results,[2] or in finding procedures for circumventing the simultaneity problem in estimation.[3] The ideal procedure is to specify a behavioral model which fully takes account of the simultaneity in determining input and output levels,[4] but lack of data frequently prevents use of this approach. For estimation of a behavioral model, data on input and output prices are normally required in addition to data on input and output levels, and price data are not available for many applied studies.

Fortunately, both price and quantity data are available for the present study. Estimation of a behavioral model is both feasible and desirable. Rather than being a hindrance simultaneity is turned to advantage because the additional relationships derived from the behavioral model enable more precise parameter estimates to be obtained. Also, as was emphasized in the preceding chapter, behaviroal assumptions are implicit in the concept of supply. The quality of the estimates obtained with the behavioral model gives an indication of the accuracy of those assumptions.

The deterministic equations of the behavioral model were derived in the preceding chapter, and a description of the data was provided there. The first part of this chapter contains a discussion of the way in which the error terms may be incorporated in the model and the assumptions made about the distribution of those errors. This is followed by a brief discussion of estimation procedures and properties of the parameter and standard error estimates obtained using those procedures. The bulk of the chapter is then devoted to a presentation and interpretation of the results obtained in estimating the model.

STOCHASTIC SPECIFICATION

The specification of the error terms in the structural equations is designed to satisfy certain a priori expectations about the properties of the error structure while at the same time yielding equations which are tractable from the standpoint of estimation. On a priori grounds one would expect that producers' errors in determining desired input and output levels are likely to be proportional to those levels as are fluctuations caused by nature, by input suppliers, etc. Also, errors in estimating the amount actually discovered are likely to be proportional to the magnitude of the discovery. For these reasons, the structural equations were assumed to have multiplicative errors. When the equations are written in logarithmic form, the errors are additive. The stochastic specification of the structural model in (5–14) is then:

(a) $\dfrac{1}{d} \text{Log} \left[a Y^d + (1-a) G^d \right] = \text{Log} \left[AF^{m+\beta} \left(\dfrac{C}{m} \right)^{\beta} e^{gW} \right] + u_1$

(b) $\text{Log} \left(\dfrac{Y}{G} \right) = \dfrac{1}{d-1} \left[\text{Log}(1-a)P_o - \text{Log}(a)P_g \right] + u_2$ (6–1)

(c) $\text{Log}(CF) - \text{Log}(P_o Y) = \text{Log} \left[m + m \left(\dfrac{1-a}{a} \right) \left(\dfrac{G}{Y} \right)^d \right] + u_3$

Equations (a) and (c) above are clearly nonlinear in the variables and parameters so standard linear regression methods cannot be applied. The method of maximum likelihood can, however, be applied to obtain estimates of the parameters. This procedure requires the assumption of a specific form for the distribution of the error terms. For estimates to be reported below, the errors were assumed to be normally distributed and serially uncorrelated with variance-covariance matrix Σ.

The joint distribution of the errors is then the product of the T independent, identical normal distributions corresponding to the T observations:

$$P(U) = \frac{1}{(2\pi)^{\frac{NT}{2}} |\Sigma|^{\frac{N}{2}}} e^{\frac{-1}{2} \sum_{t=1}^{T} U_t{}' \Sigma^{-1} U_t}$$ (6–2)

where $U_t = [u_{1t}, u_{2t}, u_{3t}]$ is the vector of errors at time t and $N = 3$ is the number of equations.

Since the true error terms are unobservable, the distribution in (6–2) does not contain any empirically observable quantities. However, equations (6–1) express the endogenous variables as functions of the parameters, the exogenous variables, and the error terms. These equations and the distribution in (6–2) can be used to derive the conditional distribution of the endogenous variables given the exogenous variables and the parameters. The resulting distribution, termed the likelihood function, written in logarithmic form is: [5]

$$\text{Log}(L) = -\frac{NT}{2} \text{Log}(2\pi) - \frac{N}{2} \text{Log}|\Sigma|$$

$$-\frac{1}{2} \sum_{t=1}^{T} U_t' \Sigma^{-1} U_t + \sum_{t=1}^{T} \log \left| \frac{\partial U_t}{\partial Z_t} \right| \qquad (6-3)$$

where $U_t = U_t(Z_t, X_t, \theta)$, and $Z_t = [Y_t, G_t, F_t]$ is the vector of endogenous variables, $X_t = [P_{ot}, P_{gt}, C_t, W_t]$ is the vector of exogenous variables, $\left| \frac{\partial U_t}{\partial Z_t} \right|$ is the absolute value of the determinant of the Jacobian of the errors with respect to the endogenous variables, and $\theta = [A, d, a, m, g, \beta, b, h]$ is the vector of parameters to be estimated. The equations in (6-1) are solved for the error terms and the resulting expressions are substituted where U_t appears in equation (6-3). Equation (6-3) is thus a function of observable quantities—the endogenous and exogenous variables, and of the parameters to be estimated.

ESTIMATION PROCEDURES AND PROPERTIES OF ESTIMATORS

Parameter estimates were obtained using the principle of maximum likelihood, which, as the name implies, involves finding the set of parameter values for which the likelihood function achieves a maximum. Since equation (6-3) is a highly nonlinear function of the parameters and the variables, finding the parameter values for which this function achieves a maximum is a non-trivial problem. Fortunately, numerical optimization procedures are available for solving this problem. The two numerical optimization procedures used in this study were Powell's conjugate gradient method and the quadratic hill climbing method of Goldfeld and Quandt.[6] The latter procedure gave consistently and often markedly better results in terms of the value of the likelihood function at the optimum. Therefore, all results reported below were obtained with the quadratic hill climbing method.

While they are not necessarily unbiased, parameter estimates which yield a global maximum of a likelihood function satisfying certain regularity conditions are consistent and asymptotically efficient.[7] Estimates of the standard errors of the coefficients are obtained by taking the square roots of the diagonal elements of the negative inverse of the matrix of second partial derivatives of the log of the likelihood function. If the maximum likelihood estimators are sufficient statistics for the parameters, standard errors so obtained will be consistents.[8] [a]

[a]The likelihood function in this study is an extremely complex function of the parameters so the sufficiency and regularity conditions noted above could not be established analytically. However, a Monte Carlo study of a similar model revealed that parameter estimates from the structural model were unbiased as were the numerically computed standard errors. A more extensive discussion is presented in Appendix A.

Treatment of Prices as Exogenous Variables

It may appear at first glance that input and output prices should also be treated as endogenous in the estimation of the model. If this were the case, the model would not be fully identified. Additional relationships would have to be specified, or estimation of the model would have to be abandoned. Prices are, however, exogenous to the present model for the following reasons.

The unit cost of drilling is dependent on the cost of labor, materials, fuel, etc. The proportion of these commodities used in petroleum exploration is a small component of the national consumption of those commodities. Their unit prices can thus be considered exogenous to the petroleum exploration industry. In the period since World War II, production from new oil and gas discoveries during the discovery year has been a small proportion of total production. Consequently, prices of oil and natural gas are little affected by the volume of new discoveries in the discovery year. It follows that current year prices are exogenous from the standpoint of exploration.

If the analysis in the preceding chapter is correct, petroleum exploration firms base their decisions on expectations of prices in the future and not on prices in the discovery year alone. In fact, most of the decisions determining drilling will have been made before prices in the discovery year can be observed. While current prices may be an accurate reflection of producers' expectations of future prices, it is more likely that expectations are based on the observed pattern of prices over several years. Current prices were used in preliminary estimation of the model, but subsequent estimates were obtained using a geometric weighting of past prices as a measure of price expectations. For those estimates, expected prices are exogenous by definition since they are a weighting of prices determined in the years preceding discovery.

ESTIMATION OF THE MODEL

The parameters to be estimated are A, d, a, m, g, β, b, and h. For estimates of these parameters to be reasonable, they must satisfy several restrictions. Those restrictions are as follows:

1. The outputs cannot be negative, $A \geqslant 0$.
2. The second order conditions for profit maximization require that the isoquants be concave to the origin, $d \geqslant 1$.
3. The standard normalization of the CET function requires $0 \leqslant a \leqslant 1$.
4. The second order profit maximization conditions require non-negative and non-increasing returns to scale, $0 \leqslant m \leqslant 1$.
5. The elasticity of supply of oil-bearing land should be non-negative, $\beta \geqslant 0$.

While a complete explanation of the final results is provided in the concluding portion of this chapter, a brief economic interpretation of the parameters at this point will facilitate understanding of the following discussion of the econometric estimates. The parameters of greatest interest are m, β, d, and g. Together, parameters m and β determine the price elasticity of supply of new discoveries of crude oil and natrual gas. When the prices of crude oil and natural gas increase proportionately by, say, one percent, discoveries of crude oil and natural gas increase by $\frac{(m+\beta)}{(1-m-\beta)}$ percent. The price elasticity of new oil discoveries when both prices move together is then $\frac{(m+\beta)}{(1-m-\beta)}$.

The elasticity of transformation, $s = \frac{1}{d-1}$, measures the percentage change in the ratio of discoveries in response to a given change in the ratio of prices of the outputs when the level of inputs is held constant. If the ratio of the price of oil discoveries to the price of gas discoveries changes by 1 percent, the ratio of the quantities discovered will change by s percent if input levels are held constant. Parameter g measures the net effects of two opposing influences: technical progress which tends to increase the productivity of exploratory drilling, and exhaustion of undiscovered deposits which tends to reduce the productivity of drilling. If g is positive, the former effect dominates; if negative, the latter dominates. Parameters A and a are essentially scaling parameters and not of particularly great interest.

Structural Estimates

Due to the highly nonlinear character of the equations, the simplest possible form of the model was estimated first. Generalizations were then made on the basis of the a priori specification of the model and the nature of the results obtained at each step. While the discussion could be limited to presentation of the final results, it seems best to present the sequence of estimates which was obtained. This will give an idea of the effects of various restrictions of the model and will indicate why certain generalizations of the model were accepted or rejected. The decision to introduce the elasticity of supply of oil-bearing land was not made until a fairly late stage in the analysis of the model so the restriction $\beta = 0$ is incorporated in the early results.

All results reported in this section are Full Information Maximum Likelihood estimates of the three equations of the structural model. The purpose of the estimates report in table 6-1 is to test whether the CET production function is an unduly restrictive functional form. The alternative against which the CET is tested is the more general function in equation (5-12) of the preceding chapter. The value of the log of the likelihood function is reported at the bottom of the table,[b] and the ratios of the coefficients to their estimated standard errors are reported in parentheses.

---

[b]An additive constant $[-(NT/2)\text{Log}\,(2\pi) - NT/2]$ has been excluded from the reported value of the log of the likelihood function since it is irrelevant for both hypothesis testing and estimation. Here T is the number of observations and N is the number of equations.

The estimates in equation (6-1-1) were obtained using the CET model, equations (6-1).[c] These equations incorporate the restrictions $d_1 = d_2$, $m_1 = m_2$, and $g_1 = g_2$. As indicated in Chapter 5, these restrictions imply equal supply elasticities for both crude oil and natural gas and equal shifting of the supply equations as a result of either technical progress or exhaustion. Equation (6-1-2) allows for differing elasticities of output, $m_1 \neq m_2$, but retains the other constraints. All three constraints were eliminated for the estimates in equation (6-1-3). The results for the latter two equations were obtained by estimating equation (5-12) of the preceding chapter along with the associated first order conditions for that equation.

The appropriate test of the significance of relaxing these constraints is a likelihood ratio test. As parameter restrictions are relaxed, the value of the likelihood function at its maximum increases. The change in the log of the likelihood function, when multiplied by -2, is distributed asymptotically as a Chi-square. The number of degrees of freedom is equal to the difference in the number of parameters in the two equations being compared.

In applying this test to the first two equations in table 6-1, it is found that the difference in the likelihood functions is -1.33 which, when multiplied by -2, yields a Chi-square value of 2.76. The second equation contains one more parameter than the first so the number of degrees of freedom is one. A Chi-square value of 2.76 with one degree of freedom is not significant at the five percent level. Thus, the null hypothesis that $m_1 = m_2$ cannot be rejected. Likewise, a test of the first and third equations of table 6-1 reveals that the null hypothesis $m_1 = m_2$, and $g_1 = g_2$ cannot be rejected at the 5 percent level. On this basis, the CET function does not appear to be unduly restrictive. This conclusion must be considered tentative, however, because none of the results in table 6-1 is entirely satisfactory.

Generalizations of the CET Model

The reason that the results in table 6-1 are unsatisfactory is not the set of parameter estimates, but rather the pattern of residuals from the estimated equations. The residuals in these equations display distinct trends, not runs as one would expect if the problem were simply serial correlation. The estimates in table 6-2 are generalizations of the model which will serve to identify the cause of the troublesome residual pattern. All of these estimates are based on the CET production function. Equation (6-1-1) is repeated as (6-2-1) for convenience of comparison.

As explained in connection with equation (5-29) in the preceding chapter, the constant $b = \dfrac{.41b_2}{b_1}$ has been introduced both to convert the IPAA index to cost per foot and to reflect the fact that a proportion of revenues from oil

---

[c]References in the text to equations in the tables contain chapter number, table number, and equation number. Hence, (6-1-1) refers to equation (1) of table 6-1.

Table 6-1. Tests of Restrictions on the CET Function

	Equation		
Coef.	(1)	(2)	(3)
a	.645 (68.4)	.091 (.66)	.570 (.47)
d_1	1.46 (9.36)	1.49 (9.73)	1.16 (2.44)
d_2	1.46 (9.36)	1.49 (9.73)	2.26 (5.61)
B	708. (.77)	435. (.92)	454. (.61)
m_1	.355 (7.16)	.279 (4.67)	.265 (3.93)
m_2	.355 (7.16)	.458 (6.93)	.540 (6.96)
g_1	.019 (.19)	-.051 (.53)	-.072 (.37)
g_2	.019 (.19)	-.051 (.53)	.063 (.46)
Log(L)	-532.70	-531.32	-529.68

production will go to pay development and production costs. Based on data reported in Appendix B, it was estimated that $b_1 \cong .41$ and $b_2 \cong 8.40$. For the estimates reported in table 6-1, b was thus set at 8.40. In equation (6-2-2), b was estimated along with the remaining parameters of the model. The estimate of 8.69 in equation (6-2-2) is remarkably close to the value used previously suggesting that the estimates based on prior information were quite good. This change did not affect the trend in the residuals noted earlier, nor did it affect the remaining parameter values or the value of the likelihood function to a significant degree.

As explained in connection with equations (5-23) and (5-29) of Chapter 5, there may also be a trend in the share of revenues attributable to exploration or a trend in costs relative to the IPAA cost index. Since prices in the equations are divided by the cost variable, both of these trends will be reflected in parameter h in equation (5-23) of Chapter 5. Both of the trends affect parameter h in the same direction; h should be positive. While the estimate of h in equation (6-2-3) is positive as expected, a value of 10 percent is somewhat large to be attrubutable to the sources identified thus far.

Whatever the reason, it is clear from the very large change in the value of the likelihood function that the introduction of h has greatly improved the fit of the equations. It is also interesting to note that neither d nor m has changed in value by a significant amount. The constant b no longer has a clear meaning since the trend term shifts the cost index thereby altering the implied value of b. The value of the parameter g has now increased considerably to about 5.9 percent.[d] There is clearly an interaction between the trends measured by these parameters. This interaction will be explored further below, but first some additional generalizations of the model will be introduced in table 6-2.

Table 6-2.   Generalizations of the CET Model

Coef.	Equation						
	(1)	(2)	(3)	(4)	(5)	(6)	(7)
a	.645 (68.4)	.645 (68.0)	.645 (66.3)	.670 (71.2)	.668 (53.5)	.666 (63.4)	.666 (67.5)
d	1.46 (9.36)	1.46 (6.21)	1.51 (15.8)	1.40 (20.1)	1.30 (20.1)	1.37 (24.5)	1.32 (19.0)
A	89.6 (.77)	80.2 (.57)	164.4 (.74)	6.15 (1.89)	6.58 (1.48)	6.40 (1.84)	2.11 (.93)
m	.355 (7.16)	.363 (2.30)	.364 (2.87)	.667 (14.3)	.657 (11.8)	.673 (14.9)	.430 (2.93)
g	.019 (.19)	.014 (.20)	−.586 (9.23)	−.599 (9.46)	−.559 (5.64)		
b		8.69 (2.20)	27.65 (2.72)	40.9 (9.49)	37.1 (6.14)	40.4 (6.23)	25.8 (2.38)
h			10.06 (11.1)	8.12 (9.52)	7.47 (5.21)		
g_s						−.524 (5.75)	−.672 (4.48)
g_d						.231 (3.87)	.082 (.45)
q				.219 (4.50)	.234 (3.67)	.318 (4.97)	.295 (3.84)
β							.31 (1.48)
r_1					.45 (3.23)	.734 (2.45)	.341 (2.00)
r_2					.48 (3.54)	.26 (1.66)	.28 (1.65)
r_3					.50 (3.51)	.42 (2.64)	.38 (2.13)
Log(L)	−532.70	−532.70	−510.00	−496.26	−488.54	−488.80	−488.33

Thus far, current price has been used in the estimation. From a theoretical standpoint, however, it is clear that expected price should be used, and that current and expected price are not likely to coincide. It has been emphasized

[d]In the estimation, the time variable was incremented by 0.01 per period so h can be interpreted directly as an annual percentage change. The unit of measure for cumulative drilling, W, is 100,000 wells. Since approximately 10,000 wells are drilled per year, W is incremented by about 0.1 per year. Therefore, g must be multiplied by 10 to obtain an approximate annual percentage change.

previously that this is a non-trivial issue in the case of oil exploration since the value of a discovery depends on price and cost conditions throughout the life of the reservoir. For this reason, a geometric weighting of past prices was introduced as a means of obtaining an expected price series.[9] Thus:

$$P_t^e = q \sum_{i=0}^{\infty} (1-q)^i P_{t-i-1} , \qquad (6-4)$$

or,

$$P_t^e = q \sum_{i=0}^{t-1} (1-q)^i P_{t-i-1} + (1-q)^t P_0^e . \qquad (6-5)$$

In the above equation, the history of prices prior to the time of the initial observation is summarized as P_0^e. This term can be treated as a parameter to be estimated, but a statistically significant estimate of the parameter cannot be obtained.[10] Asymptotically, the estimates of the remaining parameters are not affected by the treatment of P_0^e so one could simply set it equal to zero. A more appealing procedure for small samples is to assume that observed price at time one equals expected price at time zero. This procedure was adopted for all three prices. Also the value of q was assumed to be the same for all three prices to minimize the number of parameters to be estimated.

The result obtained when expected prices are introduced is equation (6-2-4). Once again, a very substantial and statistically significant improvement in the likelihood function was obtained. As would be expected, both the elasticity of output, *m,* and the elasticity of transformation, $s = \dfrac{1}{d-1}$, have increased. These elasticities may be interpreted as long-run responses to changes in prices and costs whereas the previous results, based on current prices, are short run in nature. The desire to obtain long-run responses to changes in price and cost conditions motivated the initial development of the geometric weighting procedure as a means of estimating expected price.[11] The longer the planning horizon for production of a commodity, the greater the divergence to be expected between short-run and long-run responses. The planning horizon for crude oil production is very long indeed, and the substantial difference between the estimated short- and long-run elasticities is thus in accord with theoretical expectations.

A Digression to Further Test the
CET Function

The changes adopted in equations (6-2-2) through (6-2-4) have eliminated the trend in the residuals noted in connection with the estimates in table 6-1. Because of the trend in the residuals, the tests of the restrictiveness of the CET function reported in that table were somewhat suspect. Also, despite the fact that the more general specification in equations (6-1-2) and (6-1-3) did not yield a significant improvement in the likelihood function, the differences in the estimates of m_1 relative to m_2 and d_1 relative to d_2 were quite large. For these reasons, a further test of the restrictiveness of the CET function was conducted at this point in the analysis. Equation (6-2-4) was generalized to allow differing elasticities of output for oil and natural gas using the function in equation (5-12). The result of the test is reported as equation (6-3-2) of table 6-3. Equation (6-2-4) is reproduced as (6-3-1) for comparison. Equation (6-3-2) yields a trivial improvement of the likelihood function relative to the preceding equation, and the estimated values of m_1 and m_2 are remarkably close together. This verifies the conclusion adopted tentatively on the basis of the results in table 6-1 that the CET function is sufficiently general as a representation of the technology of crude oil exploration.

Table 6-3.   A Further Test of Restrictions on the CET Function

		Equation	
Coef.		*(1)*	*(2)*
a		.670	.772
		(71.2)	(2.49)
d		1.40	1.40
		(20.1)	(20.0)
A		6.15	6.05
		(1.89)	(1.86)
m_1		.667	.678
		(14.3)	(11.9)
m_2		.667	.644
		(14.3)	(6.73)
g		−.599	−.599
		(9.46)	(9.62)
b		40.9	41.3
		(9.49)	(9.38)
h		8.12	8.20
		(9.52)	(9.26)
q		.219	.215
		(4.50)	(4.30)
log(L)		−496.26	−496.25

**Tests for Autocorrelation in
the Residuals**

While the residuals in equation (6-2-4) no longer have the trend pattern noted earlier, runs in the residuals are present suggesting serial correlation. Positive serial correlation in the residuals is to be expected since better than average results of exploration in a given year identify fruitful areas for drilling in the following year with a consequent increase in drilling effort and in discoveries relative to the average. Worse than average results have the reverse effect.

The most general procedure for introducing first order serial correlation is to let $U_t = RU_{t-1} + \epsilon_t$ where R is a symmetric $N X N$ matrix, N being the number of equations. To simplify computation and reduce the number of parameters to be estimated, R was assumed to be a diagonal matrix with elements r_1, r_2, and r_3 in equations (1a), (1b), and (1c), respectively. The assumption that R is diagonal implies that the error in a particular equation in period $(t-1)$ may affect the error in that same equation in period t, but not the errors in the other equations. This is not a particularly stringent assumption. The period one error term in each equation is assumed to have mean zero and the same variance as the remaining error terms of that equation.[12] A priori restrictions are that all three serial correlation coefficients lie in the range from -1 to 1.

The estimated values for these parameters in equation (6-2-5) are 0.45, 0.48, and 0.50; all are statistically significant at the 5 percent level. As expected, there is positive serial correlation in all three equations. Again a large and statistically significant improvement in the likelihood function is obtained. The results in equation (6-2-5) meet all of the a priori parameter restrictions, and all of the parameters except A are highly significant statistically. Parameter A is not of particularly great interest anyway since it is the constant term in the production function.

Analysis of Trends in the Supply Equations

It was noted earlier in connection with equation (6-2-4) that the value of h is larger than was expected, and that there appears to be some interdependence between parameters g and h. Though the values of both parameters are smaller in absolute value in equation (6-2-5), the pattern is essentially the same. The roles of these parameters are most easily understood in the context of the supply equations for crude oil and natural gas discoveries and the derived demand equation for drilling. Those equations, the reduced form, are obtained by algebraically solving the structural model, equation (5-14) of Chapter 5 for Y, G, and F. The resulting equations are the following:

(a)
$$Y = \left[A\left(\frac{m}{b}\right)^{-1} m_a \frac{1}{d}\right]^{\frac{1}{1-m-\beta}} \left(\frac{P_o^{m+\beta}}{C_I^m}\right)^{\frac{1}{1-m-\beta}} \times$$
$$\left[1+\left(\frac{1}{1-a}\right)^{\frac{1}{d-1}}\left(\frac{P_g}{P_o}\right)^{\frac{d}{d-1}}\right]^{\frac{d(m+\beta-1)}{d(1-m-\beta)}} e^{\left(\frac{1}{1-m}\right)(gW+mht)}$$

(b)
$$G = \left[A\left(\frac{m}{b}\right)^m (1-a)^{\frac{1}{d}}\right]^{\frac{1}{1-m-\beta}} \left(\frac{P_g^{m+\beta}}{C_I^m}\right)^{\frac{1}{1-m-\beta}} \times \qquad (6-6)$$
$$\left[\left(\frac{1-a}{a}\right)^{\frac{1}{d-1}}\left(\frac{P_o}{P_g}\right)^{\frac{d}{d-1}} + 1\right]^{\frac{d(m+\beta)-1}{d(1-m-\beta)}} e^{\left(\frac{1}{1-m}\right)(gW+mht)}$$

(c)

$$F = \left\{ \frac{A\left(\frac{m}{b}\right)^{1-\beta}}{[a(1-a)]^{\frac{1}{d}}} \right\}^{\frac{1}{1-m-\beta}} C_I^{-\left(\frac{1-\beta}{1-m-\beta}\right)} \times$$

$$\left[(1-a)^{\frac{1}{d-1}} P_o^{\frac{d}{d-1}} + a^{\frac{1}{d-1}} P_g^{\frac{d}{d-1}} \right]^{\frac{d-1}{d(1-m-\beta)}} e^{\left(\frac{1}{1-m}\right)} (gW+ht)$$

The trend in the supply equations for Y and G is $\left(\frac{1}{1-m}\right)(gW + mht)$. The trend in the equation for F is $\left(\frac{1}{1-m}\right)(gW + ht)$. The trends in the equations are different because of the appearance of mht in the first expression but only ht in the second. Four separate factors which might cause differential shifting of these equations have been identified previously: (1) progress in the application of scientific exploration techniques (geological and geophysical methods, etc.) may increase the probability that a given well will be successful; (2) technical progress in drilling techniques which reduce the real cost of drilling may not be fully reflected by the cost index being used; (3) depletion of discovery opportunities will tend to reduce the productivity of exploratory drilling; and (4) the share of revenues from production which are attributable to exploration may be increasing over time.

The differential rates of shifting of the supply equations relative to the derived demand equation can be measured much more simply by eliminating g and h and introducing a trend parameter, g_s, in the supply equations for Y and G and a second trend parameter, g_d, in the equation for F. Also, since W and t are highly correlated, a further simplification can be achieved by using W as the trend variable in all three equations. Thus, $g_s W$ takes the place of $\left(\frac{1}{1-m}\right)(gW + mht)$ in equations (6-6a) and (6-6b), and $g_d W$ takes the place of $\left(\frac{1}{1-m}\right)(gW + ht)$ in equation (6-6c).

Parameters g_s and g_d are incorporated in the structural equations by setting g and h equal to zero and by placing $Ye^{g_s W}$, $Ge^{g_s W}$, and $Fe^{g_d W}$ where Y, G, and F appear in those equations. The structural estimates obtained when these changes are adopted are reported as equation (6-2-6). The changes in the parameter estimates relative to (6-2-5) are due to the use of cumulative wells, W, as the trend variable in all three equations where trend variables W and t had appeared previously. The estimates of g_s and g_d imply that the supply equations are shifting to the left approximately 5.2 percent per year while the derived demand equation for F is shifting to the right by about 2.3 percent per year.

The Elasticity of Supply of
Oil-Bearing Land

The final parameter to be introduced in the equations is β, the elasticity of supply of oil-bearing land. The estimates obtained when this parameter is introduced are reported in equation $(6-2-7)$.[e] Though β is positive, as expected, it is not statistically significant. Not surprisingly, the estimate of m has dropped considerably. The estimates in equation $(6-2-7)$ imply that the elasticity of supply of new discoveries is attributable partially to the elasticity of supply of oil-bearing land, β, and partially to the substitution of other inputs for oil-bearing land, reflected in m. When β is set equal to zero, parameter m picks up both effects.

Failure to obtain a statistically significant estimate of β is disappointing but not surprising. Since the price and quantity of oil-bearing land are not observable, two equations had to be eliminated from the model in the course of the development in Chapter 5. As a result, the value of β is estimated indirectly on the basis of movements in the other variables. Faced with this lack of price and quantity data for oil-bearing land and relatively modest movements in the remaining price variables, one should not expect a terribly precise estimate of β to be obtained. Unfortunately, this parameter plays more than a minor role in determining the properties of the supply equation as the interpretation of the results presented in a later section of this chapter will indicate.

At this point, twelve parameters have been estimated with only twenty-two observations of data. The precision of the parameter estimates is, on the whole, quite good. Statistically significant estimates of all the parameters of interest except β have been obtained. The parameters are all of the correct sign, they satisfy all of the a priori restrictions,[f] and all are of a reasonable order of magnitude. An economic interpretation of these results will be provided below, but first some remarks on estimates obtained with an alternative specification of the error structure will be discussed.

A Further Note on Specification Of Error Structures

An important difference between linear and nonlinear models is that maximum likelihood estimates of parameters obtained from the reduced form of a nonlinear model will generally not be the same as parameter estimates obtained from the structure. In the above discussion, it has been emphasized that the reported parameter estimates are from the structural model.

[e]For all previous equations, convergence within 0.0001 of the preceding iteration for all parameters was required before the maximum likelihood estimates were accepted. Though several hundred iterations were run, this degree of precision was not achieved for equation $(6-2-7)$. Convergence within 0.001 for all parameters was achieved, and no change in the likelihood to six significant figures occurred during the last nine iterations. Also, all eigenvalues of the second partials matrix were positive indicating that the second order conditions for a maximum were satisfied. Parameters which changed from iteration to iteration were not changing monotonically but were fluctuating above and below the values reported in equation $(6-2-7)$. There is thus little doubt that these estimates are close to the optimum despite the relaxation of the criterion for convergence.

[f]During the estimation, the parameters were not permitted to go beyond the boundaries of the parameter space established by the a priori restrictions. The statement that all of the a priori restrictions are satisfied means that the likelihood function achieves a maximum in the interior of the parameter space and not at a boundary.

In order to determine the properties of parameter estimates from the two alternative specifications of the error structure, two tests were conducted. First, a Monte Carlo study of a similar model was conducted to determine the qualitative properties of parameter estimates obtained with the structural and reduced form specifications. The effects of misspecification were also tested by estimating the structure when the errors were generated in the reduced form and vice versa.[13]

The second step was to estimate the reduced form of the model of petroleum discoveries developed in this and the preceding chapters. Those reduced form estimates are reported in Appendix A. In that appendix, the properties of the structural and reduced form results in the model of petroleum discoveries are compared to the properties of the results obtained from the Monte Carlo study.

Two conclusions emerge from that comparison. First, the reduced form estimates of the parameters of the petroleum discoveries model are quantitatively in very close agreement with the structural estimates reported in this chapter. The only exception is the value of d which is on the order of 1.8 in the reduced form model; this may be compared with the value of 1.37 reported in equation (6-2-6) above. Second, the results of the petroleum discoveries model are qualitatively very similar to the results of the Monte Carlo study. This tends to support the accuracy of the general specification of the petroleum discoveries model developed in the preceding chapters. As regards the estimate of d, the Monte Carlo study indicates that the true value may be expected to be between the structural and reduced form estimates.

The quantitative agreement between the structural and reduced form estimates, the qualitative similarity of these empirical results with the results of the Monte Carlo simulation, and the "reasonableness" of these estimates in terms of a priori expectations and restrictions provide considerable support for the validity of the model developed in this study. An economic interpretation of the results and concluding comments are provided in the following sections.

INTERPRETATION OF THE RESULTS

The supply and derived demand equations of the model, written in logarithmic form, are as follows:

$$\text{(a)} \quad \text{Log } Y = \left(\frac{1}{1-m-\beta}\right) \text{Log}\left[A\left(\frac{m}{b}\right)^m a^{-\frac{1}{d}}\right] + \left(\frac{m+\beta}{1-m-\beta}\right) \text{Log } P_o - \left(\frac{m}{1-m-\beta}\right)$$

$$\text{Log } C_I + \frac{d(m+\beta)-1}{d(1-m-\beta)} \text{ Log } \left[1 + \left(\frac{a}{1-a}\right)^{\frac{1}{d-1}} \left(\frac{P_g}{P_o}\right)^{\frac{d}{d-1}}\right]$$

$$+ \quad g_s w$$

(b) $\quad \text{Log } G = \left(\dfrac{1}{1-m-\beta}\right) \text{Log} \left[A\left(\dfrac{m}{b}\right)^{m}(1-a)^{\frac{-1}{d}}\right] + \left(\dfrac{m+\beta}{1-m-\beta}\right) \text{Log } P_g$

$\qquad - \left(\dfrac{m}{1-m-\beta}\right) \text{Log } C_I + \dfrac{d(m+\beta)-1}{d(1-m-\beta)} \text{Log } \left[\left(\dfrac{1-a}{a}\right)^{\frac{1}{d-1}}\left(\dfrac{P_o}{P_g}\right)^{\frac{d}{d-1}} + 1\right]$

$\qquad + g_s W \hfill (6\text{-}7)$

(c) $\quad \text{Log } F = \left(\dfrac{1}{1-m-\beta}\right) \text{Log} \left\{A\left(\dfrac{m}{b}\right)^{1-\beta}[a(1-a)]^{\frac{-1}{d}}\right\} - \left(\dfrac{1-\beta}{1-m-\beta}\right) \text{Log } C_I$

$\qquad + \dfrac{d-1}{d(1-m-\beta)} \text{Log } \left[(1-a)^{\frac{1}{d-1}} P_o^{\frac{d}{d-1}} + a^{\frac{1}{d-1}} P_g^{\frac{d}{d-1}}\right] + g_d W$

Upon substituting the parameter estimates from equation (6-2-7), these equations become:

(a) $\quad \text{Log } Y = -2.768 + 2.869 \text{ Log } P_o - 1.678 \text{ Log } C_I$
$\qquad\qquad (1.03) \quad (2.74) \qquad\qquad (2.90)$

$\qquad -.0685 \text{ Log } \left[1. + 8.94\left(\dfrac{P_g}{P_o}\right)^{\frac{4.15}{(4.15)}}\right] - .672 \ W$
$\qquad\quad (.53) \qquad\qquad (1.79) \qquad\qquad\quad (4.48)$

(b) $\quad \text{Log } G = -.728 + 2.869 \text{ Log } P_g - 1.678 \text{ Log } C_I \hfill (6\text{-}8)$
$\qquad\qquad (.70) \quad (2.74) \qquad\qquad (2.90)$

$\qquad -.0685 \text{ Log } \left[.112\left(\dfrac{P_o}{P_g}\right)^{\frac{4.15}{(4.15)}} + 1\right] - .672 \ W$
$\qquad\quad (.53) \qquad\quad (1.28) \qquad\qquad\qquad (4.48)$

(c) $\quad \text{Log } F = -3.62 - 2.678 \text{ Log } C_I + .931 \text{ Log } \left[.031 \ P_o^{\frac{4.15}{(4.15)}}\right.$
$\qquad\qquad\qquad\qquad (4.53) \qquad\qquad (5.08) \qquad\quad (1.33)$

$\qquad \left. + .279 \ P_g^{\frac{4.15}{(4.15)}}\right] + .082 \ W$
$\qquad\quad (4.4) \qquad\qquad\quad (.45)$

The ratios of the coefficients to their estimated standard errors are reported in parentheses. In these equations and in the subsequent discussion, P_o, P_g, and C_I are expected prices and costs, unless otherwise indicated.

Consider first the supply function for crude oil discoveries, equation (6-8a). When the expected prices of crude oil and natural gas change by the same proportion, the term in square brackets in that equation is a constant. Thus, other things constant, a 1 percent increase in both (expected) prices will yield a 2.87 percent increase in discoveries.[g] A 1 percent increase in costs will result in a 1.68 percent decrease in discoveries. A change in costs has a relatively smaller effect than a change in price because the cost increase will induce a substitution of land for drilling inputs. This factor substitution partially offsets the effect of the cost increase.

When prices do not move together, the relative quantities of oil and gas discovered will change as firms focus more effort on finding the output whose price has increased. The elasticity of transformation, $s = (\frac{1}{d-1})$, measures the percentage change in the ratio of oil to gas discoveries in response to a given percentage change in the ratio of their prices when input levels are held constant. The estimates in equation (6-2-7) imply that $s = 3.15$ which, in turn, implies a substantial response of the discovery ratio to a change in the price ratio.[h]

Of course, when the price ratio changes as a result of an increase in the price of one of the outputs, the level of inputs will not remain constant. Figure 6-1 illustrates the response of discoveries when one or both prices change. Suppose that the initial equilibrium is point A with discovery levels Y_0 and G_0. If both prices increase proportionately, drilling will increase as indicated by equation (6-8c) and outputs of both products will increase to point B with discoveries Y_1 and G_1. The amount of the increase is measured by the elasticity $(\frac{m+\beta}{1-m-\beta})$. The ratio of the outputs will be the same as at point A.

If the price of crude oil increases by the same proportion as before, but the price of natural gas does not change, drilling effort will increase by a smaller amount than in the previous case. Consequently, the transformation frontier will not shift outward as far. In response to a change in the price ratio, the output ratio will change to some point, C, on the new transformation frontier with outputs Y_2 and G_2. If the coefficient $[\frac{d(m+\beta)-1}{d(1-m-\beta)}]$ is positive, Y_2 will exceed

---

[g]Note that even though β was not statistically significant, the elasticity $(m + \beta)/(1-m-\beta)$ is significant. As indicated earlier, an increase in the price of output induces an increase in the use of oil-bearing land and a substitution of other inputs for land as the price of land is bid up. Due to the relatively modest price movements during the period in question and a lack of data on oil-bearing land, these effects are difficult to distinguish empirically. Their combined effect is measured with greater accuracy.

[h]The reduced form estimate of d reported in Appendix A is 1.80. The discussion in that appendix indicates that the true value of d is likely to lie between the structural and reduced form estimates. Thus, a value for d of 1.6 and a corresponding elasticity of $s = 1.67$ are likely to be closer to the true values. While errors in the estimate of d of the magnitude noted above have a substantial effect on the value of s, they do not greatly affect the coefficients in the supply function. This may be verified by substituting the alternative values for d into equation (6-7).

Y_2 and G_2 will be less than G_0. If this coefficient is negative, Y_2 will be less than Y_1, and G_2 will be greater than G_0. Though the estimate of this coefficient reported in equation (6-8a) is negative, it is not significantly different from zero.

Figure 6-1. Changes in Quantities Discovered in Response to Changes in Prices

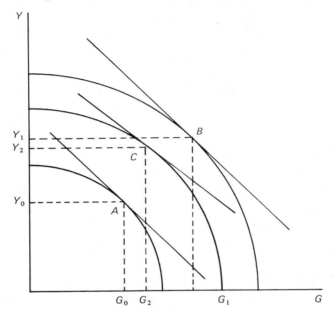

This result should not be interpreted to mean that oil and natural gas are not joint products in exploration. The fact that the two products are found and produced together is sufficient evidence that they are joint products. The above result merely says that they are neither gross complements nor gross substitutes in exploration. It is also worth noting that the model would be nonlinear even if the above coefficient were set equal to zero. The derived demand equation for F would still be nonlinear, and constraints across the equations would have to be imposed. Also, the question of whether this coefficient is or is not zero could not have been settled a priori, but only by a formal statistical test such as the one presented above.

The preceding discussion has indicated the response of discoveries to changes in expected prices and expected costs. Changes in observed prices and costs affect expected prices and costs by the adaptive adjustment mechanism specified in equation (6-5). That equation may be rewritten as follows:

$$P_t^e = (1-q)\, P_{t-1}^e + q P_{t-1} \, . \tag{6-9}$$

Expected price at time t is a weighted average of expected price at time $t-1$ and the observed price at time $t-1$. The estimated value of q in equation (6-2-7) is 0.295. This indicates that, at any time t, expected price is determined primarily by expected price in the preceding period, and adjustment of expectations in response to a change in observed price occurs relatively slowly. Substituting this value for q into equation (6-9) yields:

$$P_t^e = .705P_{t-1}^e + .295\,P_{t-1} \tag{6-10}$$

Figure 6-2 shows the response of expected price over time when the observed price changes permanently to a new level at time zero.

Figure 6-2.   Adjustment of Expected Price to Changes in Observed Price

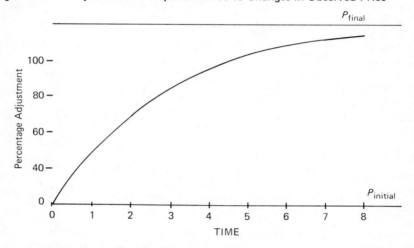

A final important factor in the supply functions is the cumulative drilling variable which measures the effects of exhaustion. As opportunities for new discoveries are depleted, the supply curve shifts to the left. The effect on the quantity discovered can be illustrated by differentiating equation (6-8a) with respect to time. For simplicity, the prices of oil and gas can be assumed to change by the same proportion. The derivative of (6-8a) is then:

$$\frac{1}{Y}\frac{dY}{dt} = 2.87\frac{1}{P_0}\frac{dP_0}{dt} - 1.68\frac{1}{C_I}\frac{dC_I}{dt} - .672\frac{dW}{dt}\ . \tag{6-11}$$

During the period in question, W has increased by an average of about 0.1 per year so $\dfrac{dW}{dt} \cong .1$. Thus, the above equation implies that the supply curve was shifting to the left about 6.7 percent per year during this period due to the effects of exhaustion.

The above equation can also be used to estimate what change in price would be required to achieve a given increase in discoveries. By rearranging this equation it is found that:

$$\frac{1}{P_0}\frac{dP_0}{dt} = .36\frac{1}{Y}\frac{dY}{dt} + .60\frac{1}{C_I}\frac{dC_I}{dt} + .23\frac{dW}{dt} . \tag{6-12}$$

This equation indicates that, to maintain discoveries at constant level (i.e., $\frac{dY}{dt} = 0$) price would have to increase by 2.3 percent per year if costs did not change.[i] Each percentage increase in costs results in a 0.6 percent increase in price, and each percent increase in the desired level of discoveries requires a 0.35 percent increase in price.

A somewhat less optimistic result is obtained if the coefficients from equation (6-2-6) are used:

$$\frac{1}{P_0}\frac{dP_0}{dt} = .485\frac{1}{Y}\frac{dY}{dt} + \frac{1}{C_I} + .254\frac{dW}{dt} . \tag{6-13}$$

The difference in the results is due to β, the elasticity of supply of oil-bearing land, which is constrained to be zero for the result in equation (6-13). Without the possibility of substituting oil-bearing land for other inputs, any increase in output must come from more intensive exploitation of the fixed quantity of land. Consequently, increases in drilling cost result in equal increases in the price of output. In addition, a relatively greater increase in price is required to achieve a given increase in the level of discoveries.

The annual change in cumulative drilling, $\frac{dW}{dt}$, will normally not be constant from period to period. If prices increase, drilling effort will increase as indicated by the derived demand equation for F, (6-8c). Since $\frac{dW}{dt}$ is proportional to the level of drilling in the preceding period, this results in an increase in the rate of shifting of the supply function. This shifting of the supply function will necessitate more rapid increases in prices to maintain a given rate of increase of discoveries.

To obtain the precise pattern of response of discoveries and drilling to change in prices and costs, the lagged response of expected price to actual price must be taken into account, and the shifting of the supply function due to exhaustion must be related to the level of drilling in the receding period. These dynamic effects can be included in a numerical simulation given a set of price and cost data for several periods.

[i]Recall that $\frac{dW}{dt} \approx .1$ during the period for which the estimates were obtained.

The results of one such simulation are indicated in figure 6-3. At $t = 1$, both the price of crude oil and the price of natural gas were doubled and were maintained at that level in all subsequent periods. The cost of drilling was increased by 6 percent per year beginning in period one.

The curves in figure 6-3 indicate the pattern of oil discoveries in subsequent periods as a percentage of oil discoveries at $t = 1$. The higher of the two curves is based on the parameter estimates in (6-2-7). The lower curve was computed with parameters from (6-2-6). The difference in the two curves illustrates once more the importance of the elasticity of supply of oil-bearing land.

Figure 6-3.   Simulated Paths of Crude Oil Discoveries

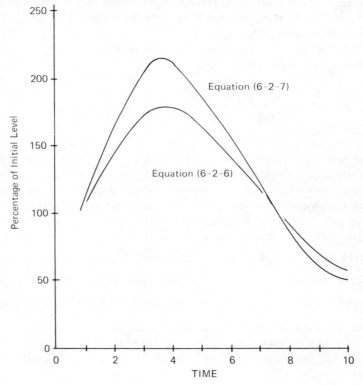

FORECASTING AND FURTHER GENERALIZATIONS
OF THE MODEL

As a rule, predictions from econometric models can be made with considerable certainty if the values of the variables remain within the bounds of the data with which the model was estimated. In time series models, it is not unusual for values of the variables to extend beyond the values observed in the sample by ten or twenty percent. However, as variables range farther and farther from the sample space, confidence declines not only because the fit of the equations and the estimates of the parameters are imperfect, but also because the specification of the model has been tested only within the confines of the sample space. Since the nature of the error, if any, in specification of the model is unknown, a formal statistical treatment of the problem is not possible.

These problems take on particular significance in the prediction of petroleum discoveries because petroleum prices have recently moved beyond the values observed within the sample period, not by 10 or 20 percent, but by more than an order of magnitude. Under these circumstances, even small errors in parameter estimates may result in sizable forecast errors. While the potential for error in predictions from an econometric model is substantial when the variables extend far beyond the sample space, any other method of prediction is surely subject to the same limitations. The model developed in this study is thus not at a disadvantage relative to other techniques of forecasting. Prudence would suggest, however, that if the model is used for predictive purposes, the relatively more pessimistic forecasts obtained from equation (6-2-6) be used at least as a supplement to forecasts from equation (6-2-7). As the data series is extended under conditions of greater variability in price, more precise estimates of the parameters (particularly β) can be obtained, and the predictive accuracy of the model should increase accordingly.

Use of the model for predictive purposes requires forecasts of prices and costs. Ideally, such price and cost forecasts should be obtained in a general equilibrium model with supply and demand functions for the various energy sources. Even in the best of circumstances, that ideal would be difficult to achieve, and the best of circumstances do not prevail. It is a considerable understatement to say that the world market for petroleum is not a competitive one, and modeling the behavior of the major foreign suppliers of crude oil would be a difficult if not impossible task.

Models of the sort developed here are useful, however, even if a general equilibrium model of the world is not available. The model can be used to predict discoveries from conventional sources under alternative assumptions about future prices, costs, taxes, and interest rates. Those predictions can serve as an aid in evaluating alternative policy proposals related to energy supplies.

The results of the analysis in this study as illustrated by equations (6-12) and (6-13) and the simulations in figure 6-3 are relatively favorable from the standpoint of developing domestic petroleum supplies. These results are based on observations in which price and cost movements have been within a relatively modest range. It is possible that the productivity of drilling will decline at an increasing rate with major increases in drilling effort. This possibility can be tested as reliable estimates of discoveries become available for years since the major price increases of 1973-74. There is neither sufficient variability in the data nor a sufficient number of observations to test the possibility with data available at the time of this study.

There are other questions which would also warrant examination. These would include an examination of the supply response of offshore relative to onshore discoveries, further tests of the substitutability between oil and natural gas discoveries, and further tests of the formation of price expectations. The results obtained in this study suggest that the model developed here should be suitable for an investigation of these questions as additional data become available.

CONCLUDING COMMENTS

The problem set forth at the outset of this study was the development of a model of the supply of crude oil and natural gas discoveries in the United States from conventional sources. The preceding chapters have in one way or another contributed to achieving that objective.

While an exhaustive analysis of technology, governmental policy, and market structure was neither possible nor desirable, these areas were examined carefully in Chapter 2. The background developed in that chapter served as a basis for much of the development in subsequent chapters. This is particularly true as regards the specification of the production function, the inclusion of policy variables in the model, and justification of behavioral assumptions used in the analysis.

An abstract model of production was developed in Chapter 3, and rough estimates of the parameters of the model were made. The analysis was motivated by the suspicion that earlier estimates resulting in the assertion of a highly elastic supply curve for crude oil were based on an incomplete model of supply. The analysis in Chapter 3 demonstrated that this was in fact the case, and the results indicated that the assertion of a highly elastic supply curve was unwarranted. While the analysis of a conventional production model served to illuminate the issues and indicate what range of price elasticities might be expected; the model did not allow for resource exhaustibility, and the procedure used for estimating the parameters of the model did not admit the application of interval estimation or hypothesis testing techniques.

As an initial step in the development of the econometric model, previously published studies were reviewed in Chapter 4. A shortcoming common to all of these models was that the estimating equations were not explicitly derived from an economic model. The studies were useful in identifying important variables and useful sources of data, but not as a guide to the specification of a model.

The principles guiding the development of the economic model in Chapter 5 were that it be a qualitatively general model based on an explicit set of behavioral assumptions. The equations of the model were derived from a production function using the assumption of competitive behavior. Resource exhaustibility was introduced via a supply function for oil-bearing land. The resulting model is a set of nonlinear simultaneous equations with parameter restrictions across the equations.

The methodology used in the empirical analysis and the results were presented in the preceding sections of this chapter. While the validity of a given economic model cannot be unequivocally confirmed or refuted by the data, the theory does provide tests of reasonableness in the form of restrictions on the signs and magnitudes of parameters of the model which, if satisfied, support the validity of the theoretical formulation. In addition, econometric methods provide established hypothesis testing procedures for evaluating the success of their application. The results presented in this chapter meet the standards for statistical significance normally imposed in econometric analysis, and the parameter estimates satisfy the a priori restrictions derived from the economic model. These properties of the estimates and the qualitative similarity of the results to those of the Monte Carlo study tend to substantiate the accuracy of the model developed in this study as a characterization of the market for the supply of petroleum discoveries from conventional sources in the United States.

The appendixes which follow contain information related to the econometric analysis presented in this chapter. Appendix A contains reduced form estimates of the parameters of the model and a comparison of those estimates of the results discussed in this chapter. A discussion of the procedures used to prepare the data series used in this analysis and a tabulation of those data series is provided in Appendix B.

Appendixes

Appendix A

Reduced Form Estimates of the Parameters of the Model

INTRODUCTION

The empirical results in Chapter 6 were obtained by estimating the structural equations of the petroleum discoveries model (Equations 6-1). There is, however, no assurance that the endogenous variables were generated by errors in the structural equations. The endogenous variables could equally well have been generated by additive errors in the reduced form of the model (Equations 6-7). Theory does not provide a basis for choosing between these competing alternatives, and there are numerous other possibilities which might be considered. Since it is customary in maximum likelihood estimation of linear models to assume that the residuals are drawn from a joint normal distribution in either the structure or the reduced form, the two alternatives considered here are a natural extension of methods applied to linear models.

In linear simultaneous equation models, the endogenous variables will have a joint normal distribution regardless of whether normally distributed disturbances are assumed to be generated in the structure or the reduced form. For most nonlinear models, including the one in this study, this is not the case. The choice of the appropriate error structure must, therefore, be determined empirically.[1]

Since there is no a priori basis for choosing between the structural and reduced form specifications, the strategy which was adopted was to estimate both. The reduced form estimates are presented below. They are compared to the structural estimates, and the qualitative properties of the two sets of estimates are compared to the results of a Monte Carlo study of a similar model.[2]

REDUCED FORM ESTIMATES

The reduced form estimates are reported in table A-1. The structural estimates from table 6-2 are reproduced as table A-2 of this appendix for purposes of comparison. The numbering of the reduced form equations corresponds to the numbering of the comparable structural estimates in table A-2. To reduce computational costs only selected tests conducted with the structure were reproduced with the reduced form.

The pattern of results obtained with the reduced form is remarkably similar to that obtained with the structural model. Pronounced trends in the residuals were obtained with the basic model in equations (A-1-1) and (A-1-2). The introduction of the trend parameter, h, equation (A-1-3) resulted in a large and highly significant increase in the value of the likelihood function. The introduction of the geometric weighting of prices in equation (A-1-4) resulted in a further significant increase in the value of the likelihood function. As with the structural model, the introduction of the expected price variable resulted in a sizable increase in m, the estimated elasticity of output with respect to drilling. Finally, the introduction of serial correlation in the error terms led to a further statistically significant improvement in the likelihood function equation (A-1-6).

Table A-1.   Reduced Form Estimates

	Equation				
Coef.	*(1)*	*(2)*	*(3)*	*(4)*	*(6)*
a	.63 (5.63)	.61 (2.78)	.64 (48.7)	.66 (26.3)	.665 (30.2)
d	3.39 (.49)	4.08 (.34)	1.67 (8.71)	2.04 (1.48)	1.80 (4.41)
A	58.1 (1.46)	28.0 (.91)	78.5 (1.08)	3.09 (8.40)	6.49 (.95)
m	.382 (7.38)	.45 (4.58)	.43 (4.98)	.72 (132.)	.67 (7.44)
g	.057 (.84)	.066 (1.07)	−.58 (9.26)	−.50 (7.10)	
b		10.13 (3.83)	30.7 (4.32)	35.47 (8.74)	41.0 (4.91)
h			9.83 (11.0)	6.63 (7.86)	
g_s					−.462 (4.80)
g_d					.312 (2.83)
q				.170 (8.34)	.214 (3.06)
r_1					.35 (1.54)
r_2					−.125 (.80)
r_3					.82 (4.99)
Log(L)	−531.72	−531.59	−508.44	−493.72	−485.04

Table A-2. Structural Estimates

Coef.				Equation			
	(1)	*(2)*	*(3)*	*(4)*	*(5)*	*(6)*	*(7)*
a	.645 (68.4)	.645 (68.0)	.645 (66.3)	.670 (71.2)	.668 (53.5)	.666 (63.4)	.666 (67.5)
d	1.46 (9.36)	1.46 (6.21)	1.51 (15.8)	1.40 (20.1)	1.30 (20.1)	1.37 (24.5)	1.32 (19.0)
A	89.6 (.77)	80.2 (.57)	164.4 (.74)	6.15 (1.89)	6.58 (1.48)	6.40 (1.84)	2.11 (.93)
m	.355 (7.16)	.363 (2.30)	.364 (2.87)	.667 (14.3)	.657 (11.8)	.673 (14.9)	.430 (2.93)
g	.019 (.19)	.014 (.20)	−.586 (9.23)	−.599 (9.46)	−.559 (5.64)		
b		8.69 (2.20)	27.65 (2.72)	40.9 (9.49)	37.1 (6.14)	40.4 (6.23)	25.8 (2.38)
h			10.06 (11.1)	8.12 (9.52)	7.47 (5.21)		
g_s						−.524 (5.75)	−.672 (4.48)
g_d						.231 (3.87)	.082 (.45)
q				.219 (4.50)	.234 (3.67)	.318 (4.97)	.295 (3.84)
β							.31 (1.48)
r_1					.45 (3.23)	.734 (2.45)	.341 (2.00)
r_2					.48 (3.54)	.26 (1.66)	.28 (1.65)
r_3					.50 (3.51)	.42 (2.64)	.38 (2.13)
Log(L)	−532.70	−532.70	−510.00	−496.26	−488.54	−488.80	−488.33

COMPARISON WITH STRUCTURAL AND MONTE CARLO RESULTS

As noted above, a Monte Carlo study of a model similar to the model developed here has been conducted. The methodology of that study will be described briefly below, and results of that study useful in the interpretation of the estimates in tables A-1 and A-2 will be summarized.

The model utilized in the Monte Carlo study contained three equations derived from a Constant Elasticity of Substitution production function using the assumption of competitive profit-maximizing behavior. The model contained five parameters corresponding to the five parameters in equation (A-1-1) of table A-1.

The effects of misspecification of the error structure were studied using the following procedure. A set of endogenous variables was generated in the structural model. The parameters were then estimated first for the correct model (the structure). Next, the same set of endogenous variables was used, but the parameters were estimated using the reduced form model (a misspecification). By replicating this procedure several times and comparing the resulting parameter estimates, the effects of misspecification when the structure was the correct model were determined. The same procedure was then repeated with endogenous variables generated in the reduced form.

The Monte Carlo study revealed that parameter and standard error estimates in the correctly specified models were quite accurate with the exception of reduced form estimates of parameter d. In contrast, misspecification of the error distribution resulted in a substantial downward bias in the estimated standard errors as well as biases in estimates of some of the parameters.

It was found that a generalized log likelihood ratio test was a highly reliable procedure for selecting the correct model. This procedure simply involves choosing the specification which yields the largest value of the likelihood function at the maximum. This procedure resulted in the choice of the correct specification for 117 out of 120 replications in the Monte Carlo study.

A comparison of the parameter estimates in equations (A-1-6) and (A-2-6) reveals that the structural and reduced form estimates agree surprisingly well. If anything, the agreement between the structural and reduced form estimates is better than the results of the Monte Carlo study, cited above, would have led one to expect. The estimates for m, the parameter of greatest interest, are identical. The estimate of d in the structural model is lower than that in the reduced form model. This is precisely the pattern for d which would have been expected on the basis of the Monte Carlo study. The reduced form estimates of d are biased upward in both the correctly specified and incorrectly specified models for that study. The structural estimates of d are biased downward in the incorrectly specified model and unbiased in the correctly specified model. In short, regardless of the correct specification of the error terms, the true value of d may be expected to fall between the upper bound estimate obtained with the reduced form and the lower bound estimate obtained with the structure.

The results in the above tables correspond most closely to the results obtained in the Monte Carlo simulation when the true model was the reduced form. In that case, the structural and reduced form estimates of m were in agreement; they were unbiased. The structural estimate of d was biased downward as was its estimated standard error. The reduced form estimate of d was biased upward, but its standard error was unbiased. The estimates of A were unbiased in both cases, and the estimates of a had moderate biases in both cases. The estimates of the trend parameter from both models were unbiased in the Monte Carlo simulation when the true model was the reduced form. Though the estimates of the trend parameters in equations (A-1-6) and (A-2-6) are not in perfect agreement, the differences are not particularly large.

The log likelihood ratio test also suggests that the reduced form is the correct specification. A comparison of the values for Log (L) in equations (A-1-6) and (A-2-6) reveals that the reduced form value of the log likelihood function is 3.7 units larger than the corresponding value for the structure.

Though the reduced form model appears to be the correct specification, the decision was made to present the structural results in Chapter 6. The reason is that a more extensive set of results was available for the structural model. While the reduced form equations can be derived explicitly for the CET model, they cannot be derived explicitly for generalizations of the model based on equation (5-12). The tests reported in tables 6-1 and 6-3 would, therefore, be extremely costly to conduct using the reduced form model. No difficulties are encountered in conducting these tests for the structural model. Since the structural and reduced form estimates for the CET model (equations A-1-6 and A-2-6) are in close agreement, the decision was made to focus on the structural results in the discussion in Chapter 6.

CONCLUSION

While the element of chance cannot be ruled out entirely, the striking similarity in the results of this econometric analysis with the results of the Monte Carlo study of a similar model appears to be more than mere coincidence. Note should be taken of the relatively greater complexity of the econometric model. Despite this qualification, the results of the Monte Carlo study can hardly help but increase one's confidence in the results of the econometric model developed in this study.

Appendix B

Data Preparation

INTRODUCTION

This appendix contains a tabulation of the data, a discussion of units of measurement, and an explanation of how certain conversion factors were estimated.

THE SHARE OF REVENUES ATTRIBUTABLE TO EXPLORATION

In equation (5-23) of Chapter 5, an estimate of the share of total crude oil revenues attributable to exploration is required. In the long run, total discounted expenditures for oil and natural gas production should equal total discounted revenues from that production under competitive conditions. Thus, ignoring the time distribution of expenditures for the various stages of production, the proportion of revenues which should be allocated to exploration should equal the proportion of exploration expenditures to total expenditures.

The Joint Association Survey lists a breakdown of industry expenditures by exploration, development, and production.[1] Unfortunately, they do not include successful exploratory wells in the exploration category. This problem can be approximately corrected by multiplying footage of successful exploratory wells by average cost per foot of exploratory wells and allocating the result to the JAS estimate of exploration costs from their estimate of development costs.

The resulting estimates of the share of expenditures in each category were presented in table 5-1 of the text. The average share of revenues in each category calculated from that table for the years 1955, 1956, and 1959 through 1965 is:[a]

Exploration	35.0
Development	32.5
Production	32.5

[a]The data in table 5-1 extend through 1968 but only data through 1965 were available to the author when these calculations were made. If the data through 1968 were used, the results would not change noticeably.

This tabulation ignores the fact that exploration, development, and production expenditures occur sequentially. Assume that exploration expenditures occur at time zero for a particular reservoir. Development expenditures occur subsequent to this but are concentrated in the years immediately after discovery and decline to a relatively low level thereafter. As a rough approximation, it may be assumed that development expenditures occur an average of two years after exploration expenditures are made. Thus, to discount development expenditures back to time zero, they should be divided by $(\frac{1}{1+r})^2$ where r is the discount rate.

Finally production expenditures occur throughout the life of the reservoir and are approximately proportional to the amount of production. Suppose production costs C dollars per unit. Then, over the life of the reservoir, expenditures will be:

$$\text{EXP} = \int_0^\infty Cq_0 e^{-Dt} dt = \frac{Cq_0}{D}. \tag{B-1}$$

This is observed expenditure. The present value of the expenditure is:

$$\text{PVEXP} = \int_0^\infty Cq_0 e^{-Dt} e^{-rt} dt = \frac{Cq_0}{(D+r)}. \tag{B-2}$$

Hence, to convert observed expenditures to present value, they must be multiplied by $\text{PVEXP}/\text{EXP} = D/(D+r)$. The average of Moody's Baa rate for the years in question is 4.70 percent, so $(\frac{1}{1+r})^2 = 0.9122$. The average crude oil decline rate for these years was 9.18 percent. Hence, $D/(D+r) = .635$. Applying these figures to the undiscounted expenditure shares listed above yields the following tabulation of expenditure shares in present value terms:

Exploration	35.0
Development	29.7
Production	20.6
Total	85.3

Thus the proportion of the present value of costs attributable to exploration is 35.0/85.3 = 0.41. This is the basis for the figure cited in the text. This calculation is obviously very rough, but the result was used only as a preliminary estimate as explained in the text. Subsequent estimates were obtained with this factor treated as a parameter of the model. Errors in the above estimate thus have no effect on the final result.

TREATMENT OF REVENUES FROM NATURAL GAS LIQUIDS

In the discussion in the text of Chapter 5 following equation (5-23), it was indicated that liquids are frequently dissolved in natural gas at the time of production. The value of a natural gas discovery should thus include revenues from the gas and from natural gas liquids (NGL) as well. The procedure for including NGL is explained below.

The American Gas Association does not report estimates of NGL discoveries. However, production data reported in the following table reveal that natural gas liquids are produced in approximately constant proportion to natural gas.

Table B-1. Ratio of Natural Gas Liquids Production to Natural Gas Production

1947	.029	1958	.030
1948	.031	1959	.031
1949	.032	1960	.033
1950	.033	1961	.034
1951	.034	1962	.034
1952	.032	1963	.035
1953	.033	1964	.035
1954	.032	1965	.034
1955	.032	1966	.034
1956	.032	1967	.035
1957	.031	1968	.036

Source: Natural gas and natural gas liquids production data taken from *Gas Facts*, published by the American Gas Association, Arlington, Va., 1971, p. 37.

Utilizing the assumption that natural gas and natural gas liquids are produced in constant proportions, the unit value of a natural gas discovery is simply the price per unit of natural gas plus the revenue from natural gas liquids per unit of natural gas produced. Since the unit value of natural gas liquids fluctuates considerably, a three-year moving average of the unit value of NGL was added to the unit price of natural gas. The resulting price series is reported in table B-2.

ESTIMATION OF A CONVERSION FACTOR FOR THE IPAA COST INDEX

Estimates of drilling cost per foot are reported in the Joint Association Survey of Industry Drilling Costs for the years 1953, 1955, 1956, and 1959 to date.[2] These data were used to convert the IPAA index to cost per foot. The average cost per foot estimated by the JAS for the above years (through 1968) was $13.47 while the average value of the IPAA index for these years was 1.60. Dividing the first figure by the second yields a value of 8.40. Since the IPAA index was 1.00 in 1947, this factor indicates an approximate cost per foot of

$8.40 in 1947. This is the factor used to convert the IPAA index to cost per foot in the preliminary estimates reported in table 6-1. For results reported in subsequent tables, this conversion factor was treated as a parameter to be estimated so the above value was not used in obtaining the final results. The simple correlation between the JAS and the IPAA data is 0.95.

To test for possible differences in trends between the JAS and IPAA data, the former set of data was regressed on the latter. If the two sets of data were proportional, the constant in the regression should not have been significant. However, as the following equation indicates, this was not the case.

$$JAS = 3.60 + 6.16IPAA \tag{B-3}$$
$$(3.67) \quad (10.2)$$

The figures in parentheses are t-statistics. The positive intercept in this equation indicates that the JAS measure did not increase as rapidly as the IPAA index. This is the basis for the statement in the text (see the discussion related to equation (5-27) of Chapter 5) that the IPAA index probably overstates the actual rate of increase in costs during this period.

CHOICE OF A DISCOUNT RATE

Moody's Baa bond rate was used as the discount rate for this analysis. It was argued in Chapter 3 that this rate is likely to underestimate the true discount rate in petroleum production. In a well-functioning capital market, however, the cost of capital in petroleum production should move in the same direction as the bond rate used in this study. This is not an entirely satisfactory argument in the present case since the discount rate enters equation (5-23) of Chapter 5 in a nonlinear fashion. Even if Moody's Baa rate differed from the true cost of capital in crude oil production only by a multiplicative constant, use of the Baa rate would not give a price index which was perfectly correlated with the price index obtained using the correct cost of capital.

Despite this drawback, the price index obtained using the Baa rate should be highly correlated with the index which would have been obtained with the true cost of capital. The price index used in this study should thus reflect the effects of important movements in the true discount rate. The alternative to using the Baa rate is to attempt to estimate a cost of capital for the petroleum industry. That is a major research project in itself, and the requirements of the present study did not justify such an undertaking. While the shortcomings of the Baa rate must be acknowledged, it is adequate for the problem at hand.

UNITS OF MEASUREMENT

A tabulation of the data used in this study is provided in Table B-2. The units of measure for this data are as follows:

Oil discoveries, Y, are measured in millions of barrels.

Natural gas discoveries, G, are measured in units of three million cubic feet. This rather unorthodox unit of measure was used in order to yield a value for parameter a which was not near either of the boundaries for that parameter. The units were thus chosen for computational convenience.

The units for drilling, F, are thousands of feet.

The units for cumulative drilling, W, are 100,000 wells.

The price of oil, P_O is in cents per barrel discovered after tax. This is calculated using equation (5-23) of Chapter 5 with $b_1 = 0.41$ and $h=0$. The value for k is estimated ultimate recovery as a proportion of original oil-in-place.

The price of natural gas, P_g is in cents per three thousand cubic feet discovered after tax. Again, equation (5-23) is used with $b_1 = 0.41$ and $k=1$. In this case, the wellhead price, P_w in equation (5-23) of Chapter 5 includes the correction for natural gas liquids discussed earlier.

The drilling index reported by the IPAA has a value of 1.00 in 1947. To prepare the cost index, C, tabulated below, the IPAA index was multiplied by $(1-T)$ to put it on an after tax basis. This result was then divided by 10. These units were chosen to yield a constant of a reasonable order of magnitude in the econometric analysis.

Table B-2.   Data Used In The Econometric Analysis

Y	G	F	W	P_O	P_g	C
5702.867	4071.350	26395.398	0.0	11.834	4.626	0.062
8614.250	2913.042	33108.852	0.068	15.633	5.401	0.062
15843.160	7705.380	34800.824	0.149	14.659	5.806	0.062
7395.824	4424.849	40172.785	0.239	14.323	6.210	0.063
6083.684	3362.802	49339.926	0.342	13.085	5.897	0.058
4824.051	5375.296	55614.297	0.460	12.783	5.940	0.060
7934.816	4301.940	60707.270	0.584	13.248	6.355	0.062
7097.355	5038.820	59578.250	0.717	13.849	7.021	0.062
5570.145	3358.375	69173.188	0.848	13.990	7.176	0.063
6256.004	6003.530	73975.250	0.998	13.853	7.027	0.066
6088.980	7243.428	69137.563	1.159	14.395	6.596	0.069
3677.539	6098.858	61457.648	1.307	13.586	6.770	0.069
2688.489	2903.617	63188.496	1.438	12.812	7.025	0.071
2963.247	4117.431	55785.141	1.570	12.566	7.526	0.073
1921.212	3923.065	54279.914	1.687	12.762	8.069	0.074
2727.323	3780.158	53509.480	1.797	13.004	8.298	0.076
1663.549	4094.570	53410.363	1.904	13.298	8.621	0.076
2966.775	2931.613	55394.223	2.011	13.650	8.744	0.081
2198.434	2914.046	49089.910	2.118	13.878	9.038	0.088
1924.253	2608.940	54817.113	2.213	13.506	8.783	0.095
2789.382	2114.038	52741.492	2.315	13.525	8.838	0.099
2798.602	1708.891	53877.770	2.406	12.580	8.418	0.099

Notes

NOTES TO CHAPTER 1
INTRODUCTION

1. Alfred E. Kahn. "The Depletion Allowance in The Context of Cartellization," *American Economic Review* 54, 4, Part I (June 1964): 286.

NOTES TO CHAPTER 2
TECHNICAL AND INSTITUTIONAL FRAMEWORK

1. Stuart E. Buckley et al., *Petroleum Conservation* (Dallas: E.J. Storm Printing Company, 1951), Chapter III.

2. R. Dana Russel, "The Evolution of Exploration Technology," in National Petroleum Council, *The Impact of New Technology on the U.S. Petroleum Industry 1946–65* (Washington: National Petroleum Council, 1967), p. 45.

3. R. Dana Russel, "Geology and Geochemistry," in National Petroleum Council, *The Impact of New Technology on the U.S. Petroleum Industry 1946–65* (Washington: National Petroleum Council, 1967), p. 71.

4. Frank J. McDonal, "Geophysics," in National Petroleum Council, *The Impact of New Technology on the U.S. Petroleum Industry 1946–65* (Washington: National Petroleum Council, 1967), p. 71.

5. Ibid., pp. 50–51 and 66–67; Interstate Oil Compact Commission, *A Study of Conservation of Oil and Gas,* Oklahoma City, by the IOCC, 1964, p. 21.

6. National Petroleum Council (NPC), *The Impact of New Technology on the U.S. Petroleum Industry 1946–65* (Washington: The National Petroleum Council, 1967).

7. Ibid., p. 6.

8. Ibid., p. 46

9. Ibid., pp. 46–47.

10. John G. McLean and Robert Wm. Haigh, *The Growth of the Integrated Oil Companies* (Boston: Division of Research, Harvard University, Graduate School of Business Administration, 1954), p. 397.

11. Figures are reported annually in the *Bulletin of The American Association of Petroleum Geologists,* June issues.

12. NPC, *Impact of New Technology,* pp. 84-85.

13. Figures reported in *World Oil,* February 15, 1959, pp. 118-19.

14. Figures reported in *Joint Association Survey of the U.S. Oil and Gas Producing Industry* sponsored by the American Petroleum Institute, the Independent Petroleum Association of America, and the Mid-continent Oil and Gas Association (Washington: The American Petroleum Institute, 1972), pp. 1 and 9.

15. NPC, *Impact of New Technology,* pp. 84-85.

16. L.H. VanDyke, "North American Drilling Activity in 1967," *Bulletin of The American Association of Petroleum Geologists* 52, 6 (June 1968): 905.

17. Buckley, *Petroleum Conservation,* pp. 175-76.

18. All oil reservoirs contain some natural gas though the proportion can vary over a wide range. The reverse, however, is not true. Reservoirs containing dry gas occur frequently. Buckley, *Petroleum Conservation,* Chapter III.

19 References for this discussion are Buckley, Ibid., Chapters III-IV; Interstate Oil Compact Commission (IOCC), *Study of Conservation,* Chapter II; Stephen L. McDonald, *Petroleum Conservation in the United States: an Economic Analysis* (Baltimore: Johns Hopkins Press, 1971), Chapter 2; and Paul T. Homan and Wallace F. Lovejoy, *Economic Aspects of Oil Conservation Regulation* (Baltimore: Johns Hopkins Press, 1967), Chapter 2.

20. Stephen L. McDonald and James W. McKie, "Petroleum Conservation in Theory and Practice," *Quarterly Journal of Economics* 76 (February 1962): 98-121.

21. IOCC, *Study of Conservation,* pp. 53 and 56; and U.S. Congress, Senate, Committee on Interior and Insular Affairs, *Trends in Oil and Gas Exploration* (Washington: U.S. Government Printing Office, 1972), Volume 1, p. 332.

22. IOCC, *Study of Conservation,* pp. 49-50; and NPC, *Impact of New Technology,* pp. 164-65.

23. R.G. Cummins and R.G. Kuller, "An Economic Model of Production and Investment for Petroleum Reservoirs," *American Economic Review* 64, 1 (March 1974): 66-79.

24. NPC, *Impact of New Technology,* pp. 9-10.

25. U.S. Congress, *Trends in Oil and Gas Exploration,* p. 327. Figures prepared by the Department of Interior.

26. NPC, *Impact of New Technology,* p. 3.

27. See Paul G. Bradley, *The Economics of Crude Petroleum Production* (Amsterdam: North-Holland Publishing Co., 1967), p. 33, and M.A. Adelman, *The World Petroleum Market* (Baltimore: The Johns Hopkins University Press, 1972), p. 27.

28. Bradley, *Economics of Crude Petroleum Production,* p. 46.

29. J.J. Arps, "Analysis of Decline Curves," *Petroleum Development and Technology,* New York: American Institute of Mining and Metallurgical Engineers, Vol. 160, 1945, p. 232. Cited by Bradley, *Economics of Crude Petroleum Production.*

30. Arps, "Analysis of Decline Curves," p. 240.

31. Bradley, *Economics of Crude Petroleum Production,* pp. 59-60.

32. Adelman, *World Petroleum Market,* p. 28.

33. VanDyke, "North American Drilling Activity in 1967," p. 913.

34. Ibid.

35. *Reserves of Crude Oil, Natural Gas Liquids, And Natural Gas in the United States and Canada and United States Productive Capacity as of December 31, 1972.* Prepared jointly by the American Petroleum Institute, The American Gas Association, and The Canadian Petroleum Association, Washington, 1973.

36. Ibid., p. 14.

37. Ibid., p. 101.

38. U.S. Bureau of Mines, *Minerals Yearbook* (Washington: U.S. Government Printing Office, Annual).

39. *Reserves of Crude Oil,* p. 19.

40. These indices are available in *Reports of the Cost Study Committee* of the Independent Petroleum Association. These reports are issued on an irregular basis by the IPAA.

41. *Joint Association Survey.*

42. NPC, *Impact of New Technology,* p. v and vi. The NPC is "an officially established industry advisory board to the Secretary of the Interior."

43. Paul T. Homan and Wallace F. Lovejoy, *Methods of Estimating Reserves of Crude Oil, Natural Gas, and Natural Gas Liquids* (Baltimore: The Johns Hopkins Press, 1965).

44. References for the material in this section are Paul T. Homan and Wallace F. Lovejoy, *Economic Aspects of Oil Conservation Regulation* (Baltimore: The Johns Hopkins Press, 1967); Stephen L. McDonald, *Petroleum Conservation in the United States: an Economic Analysis* (Baltimore: The Johns Hopkins Press, 1971); Interstate Oil Compact Commission, *A Study of Conservation of Oil and Gas in The United States* (Oklahoma City: IOCC, 1964).

45. Erich W. Zimmerman, *Conservation in the Production of Petroleum* (New Haven: Yale University Press, 1957).

46. Homan and Lovejoy, *Economic Aspects,* p. 73.

47. Melvin G. deChazeau and Alfred E. Kahn, *Integration and Competition in the Petroleum Industry* (New Haven: Yale University Press, 1959), pp. 190–98.

48. IOCC, *Study of Conservation,* p. 9.

49. Homan and Lovejoy, *Economic Aspects,* Chapter 2 and pp. 140–41.

50. *Report of the Attorney General Pursuant to Section 2 of the Joint Resolution of July 28, 1955 Consenting to an Interstate Compact to Conserve Oil and Gas* (Washington: U.S. Government Printing Office, 1957).

51. Ibid., p. 33.

52. Ibid., p. 47.

53. Morris A. Adelman, "Efficiency of Resource Use in Crude Petroleum," *Southern Economic Journal* 31, 2 (October 1964): 107.

54. An editorial in a prominent industry journal has proposed that prorationing regulations be scrapped. "Market Demand Proration Has Outlived Its Purpose," *Oil and Gas Journal* 70, 11 (March 13, 1972): 19.

55. This section relies heavily on J. Leslie Goodier, *U.S. Federal and Seacoastal State Offshore Mining Laws* (Washington, D.C.: Nautilis Press, 1972), pp. 1–11, 43–48, and 101–112.

56. Kenneth W. Dam, "Implementation of Import Quotas: The Case of Oil," *The Journal of Law and Economics* 14, 1 (April 1971): 11.

57. Ibid., p. 18

58. *Annual Statistical Review: U.S. Petroleum Industry Statistics 1956-1972* (Washington: American Petroleum Institute, April 1973), p. 5

59. *Economic Report of the President,* 1974, p. 115.

60. "Oil Import Curbs Eased for Rest of Year in Attempt to Meet Rising Fuel Demand," *Wall Street Journal,* September 19, 1972, p. 4.

61. The primary reference for this section is Stephen L. McDonald, *Federal Tax Treatment of Income From Oil and Gas* (Washington: The Brookings Institution, 1965).

62. "Conferees Clear $22.8 Billion Tax Cut; Oil Industry Taxes are Lifted $2 Billion," *Wall Street Journal,* March 27, 1975, p. 3.

63. McDonald, *Federal Tax Treatment,* p. 10.

64. Ibid., p. 14.

65. Ibid., p. 17n. Data based on a sample of firms which collectively accounted for 90 percent of all depletion deductions in the years indicated.

66. Ibid., p. 20-21.

67. Alfred E. Kahn, "The Oil Depletion Allowance in the Context of Cartelization," *American Economic Review* 54, 4 (June 1964): 287 n.

68. CONSAD Corporation, *The Economic Factors Affecting the Level of Domestic Petroleum Reserves,* in Part 4 of U.S. Congress. Committee on Ways and Means and Senate Committee on Finance, *Tax Reform Studies and Proposals,* U.S. Treasury Department (Washington: U.S. Government Printing Office, 1969), p. 5.4.

69. McDonald, *Federal Tax Treatment,* p. 17 n.

70. A particularly instructive debate on the allocation issue may be found in a series of exchanges in the *National Tax Journal* among Stephen L. McDonald, Richard Musgrave, Douglas Eldridge, and Peter O. Steiner. See Stephen L. McDonald, "Percentage Depletion and the Allocation of Resources: The Case of Oil and Gas," *National Tax Journal* 14, 4 (December 1961): 323-336. On the intertemporal effects of the depletion allowance, see the discussion of exhaustion in Chapter V.

71. "U.S. Owns Rights to Offshore Oil, Top Court Rules," *Wall Street Journal,* March 18, 1975, p. 3.

72. Goodier, *U.S. Federal and Seacoastal State Offshore Mining Laws,* pp. 1-11.

73. Public Land Law Review Commission, *One Third The Nations Land* (Washington: U.S. Government Printing Office, 1970), p. 192.

74. Paul W. MacAvoy, "The Regulation Induced Shortage of Natural Gas," *Journal of Law and Economics,* 14, 1 (April 1971): 167-99.

75. James W. McKie, "Market Structure and Uncertainty in Oil and Gas Exploration," *Quarterly Journal of Economics* 74, 4 (November 1960): 549.

NOTES TO CHAPTER 3
ECONOMIC RENTS AND THE PRICE ELASTICITY
OF U.S. PETROLEUM SUPPLY

1. Richard Mancke, "The Longrun Supply Curve of Crude Oil Produced in the United States," *Antitrust Bulletin* 15 (Winter 1970): 727-56.

2. Ibid., p. 732

3. Ibid., p. 745

4. Richard Muth, "The Derived Demand For a Productive Factor and the Industry Supply Curve," *Oxford Economic Papers,* (New Series), 16, 2 (July 1964): 227.

5. Mancke, "The Longrun Supply Curve," p. 745.

6. Marc Nerlove, "Recent Empirical Evidence on the CES and Related Production Functions," in Murray Brown (ed.), *The Theory and Empirical Analysis of Production* (New York: National Bureau of Economic Research, 1967).

7. Muth, "Derived Demand," p. 229.

8. Mancke, "The Longrun Supply Curve," p. 731.

9. Ibid., p. 741.

10. Ibid., p. 744

11. Ibid.

12. Ibid., p. 736.

13. Earnings data taken from *The Oil Producing Industry in Your State,* 1972, p. 108. Price data taken from the *Petroleum Independent,* Sept./Oct. 1971, p. 289. These two publications cite the Bureau of Labor Statistics as the source of their data.

14. See the series of articles by Edward F. Denison and by Jorgenson and Griliches in "The Measurement of Productivity," *Survey of Current Business* 52, 5, Part II (May 1972): 65–94. Figures reported in the text are from table 25, p. 89 of the article by Jorgensen and Griliches, "Issues in Growth Accounting: A Reply to Edward F. Denison."

NOTES TO CHAPTER 4
A CRITIQUE OF PRIOR
ECONOMETRIC STUDIES

1. CONSAD Corporation, *The Economic Factors Affecting the Level of Domestic Petroleum Reserves,* in Part 4 of U.S. Congress. Committe on Ways and Means and Senate Committee on Finance, *Tax Reform Studies and Proposals,* U.S. Treasury Department (Washington: Government Printing Office, 1969), p. 1.1.

2. Dale W. Jorgenson, "Capital Theory and Investment Behavior," *American Economic Review* 52 (May 1963): 247–257.

3. CONSAD, *Economic Factors,* p. 2.2.

4. Ibid., p. 3.2

5. Ibid., p. 2.2

6. Ibid., p. 7.8.

7. Ibid., p. 3.3

8. Ibid., p. 6.34.

9. Ibid., p. 6.36.

10. Ibid., p. 6.52.

11. Ibid., p. 10.6.

12. Ibid., pp. 9.6 and 9.7.

13. Franklin M. Fisher, *Supply and Costs in the U.S. Petroleum Industry: Two Econometric Studies* (Baltimore: The Johns Hopkins Press, 1964).

14. Ibid., p. 24.

15. Ibid., p. 37.

16. Edward W. Erickson, "Economic Incentives, Industrial Structure and the Supply of Crude Oil Discoveries in the U.S., 1946–58/59," Unpublished manuscript.

17. Ibid., p. 3.

18. Edward W. Erickson and Robert M. Spann, "Supply Response in a Regulated Industry: The Case of Natural Gas," *Bell Journal of Economics and Management Science* 2, 1 (Spring 1971): 94–121.

19. J.D. Khazzoom, "The FPC Staff's Econometric Model of Natural Gas Supply in the United States," *Bell Journal of Economics and Management Science* 2, 1 (Spring 1971): 51–93.

20. Ibid., p. 59.

21. Examples may be found in John Johnston, *Econometric Methods,* 2d ed. (New York: McGraw-Hill, 1971), pp. 300–303.

22. Ibid., p. 307.

23. P.W. MacAvoy and R.S. Pindyck, "Alternative Regulatory Policies for Dealing with the Natural Gas Shortage," *Bell Journal of Economics and Management Science* 4, 2 (Autumn 1973): 454–498.

24. Ibid., p. 476.

25. Ibid., p. 478.

26. R.S. Pindyck, "The Regulatory Implications of Three Alternative Econometric Supply Models of Natural Gas," *Bell Journal of Economics and Management Science* 5, 2 (Autumn 1974): 633–645.

27. Paul G. Bradley, *The Economics of Crude Petroleum Production* (Amsterdam: North-Holland Publishing Company, 1967).

28. Ibid., p. 47n.

29. See Henry Steele's statement before the Senate Subcommittee on Antitrust and Monopoly. H.B. Steele, U.S. Congress. Senate. Committee on the Judiciary. Subcommittee on Antirust and Monopoly. *Government Intervention in the Market Mechanism: The Petroleum Industry, Part 1, Economists Views* (Washington: U.S. Government Printing Office, 1969), pp. 208–233.

NOTES TO CHAPTER 5
FORMULATION OF THE ECONOMETRIC MODEL

1. Harold Hotelling, "The Economics of Exhaustible Resources," *Journal of Political Economy* 39, 2 (April 1931): 137–75.

2. Richard L. Gordon, "A Reinterpretation of the Pure Theory of Exhaustion," *Journal of Political Economy* 75 (June 1967): 274–86.

3. Frederick M. Peterson, *The Theory of Exhaustible Natural Resources: A Classical Variational Approach,* Ph.D. dissertation, Princeton University, 1972.

4. Orris C. Herfindahl, "Depletion and Economic Theory," in Mason Gaffney (ed.), *Extractive Resources and Taxation* (Madison: The University of Wisconsin Press, 1967), pp. 63–90.

5. Anthony Scott, "The Theory of the Mine Under Conditions of Certainty," in Gaffney (ed.), *Extractive Resources and Taxation,* pp. 25-62.

6. An analysis of this model may be found in Gordon, "A Reinterpretation," pp. 276-77.

7. A brief, clear exposition of the principal results of the calculus of variations is provided by Michael D. Intriligator, *Mathematical Optimization and Economic Theory* (Englewood Cliffs, N.J.: Prentice-Hall, Inc., 1971), Chapter 12. Further expositions in order of increasing complexity are R.G.D. Allen, *Mathematical Analysis for Economists* (London: Macmillan and Co., 1964), Chapter 22, and I.M. Gelfand and S.V. Fomin, *Calculus of Variations* (Englewood Cliffs, N.J.: Prentice-Hall, Inc., 1963).

8. This is the particular formulation adopted by Peterson, *Theory of Exhaustible Natural Resources,* p. III-1.

9. This is essentially the argument used sixty years ago by Lewis C. Gray, "Rent Under the Assumption of Exhaustibility," *Quarterly Journal of Economics* (Cambridge: Harvard University Press, May 1914). Reprinted in Gaffney, *Extractive Resources and Taxation,* pp. 423-446.

10. Peterson, *Theory of Exhaustible Natural Resources,* p. IV-16.

11. A good deal of effort has been directed toward a determination of the probability distribution for discoveries. See Paul G. Bradley and R.S. Uhler, "A Stochastic Model for Determining the Economic Prospects of Petroleum Exploration Over Large Regions," *Journal of the American Statistical Association* 65, 330 (June 1970): 623-30, and Gordon M. Kaufman, *Statistical Decisions and Related Techniques in Oil and Gas Exploration* (Englewood Cliffs, N.J.: Prentice-Hall, 1963).

12. A clever application of this procedure using the Cobb-Douglas production function is provided by A. Zellner, J. Kmenta, and J. Dreze, "Specification and Estimation of Cobb-Douglas Production Function Models," *Econometrica* 34 (October 1966): 784-95. Bridge has shown that the same procedure can be applied to the CES production function. J.L. Bridge, *Applied Econometrics* (Amsterdam: North-Holland Publishing Co., 1971).

13. Loosely stated, one contribution of duality theory is in determining the minimum set of conditions for which the properties of the production set and the various functional types are equivalent. A compact summary of important results is provided by W.E. Diewert "Functional Forms for Profit and Transformation Functions," *Journal of Economic Theory* 6 (1973): 284-316.

14. Ibid., p. 295.

15. L.R. Christensen, D.W. Jorgenson, and L.J. Lau, "Transcendental Logarithmic Production Frontiers," *Review of Economics and Statistics* 4, 1 (February 1973): 28-45. The terms "production frontier" or "transformation function" are often used to refer to production functions with multiple outputs. The convention adopted in this study is to refer to these as production functions regardless of the number of outputs.

16. Despite this drawback the Cobb-Douglas function was used by Klein in a study of passenger and freight transportation by railroads. He argued that the Cobb-Douglas was acceptable when output proportions were established by regulatory authority, but noted that its use could not be justified where output

proportions were chosen competitively. L.R. Klein, *A Textbook of Econometrics* (Evanston: Row, Peterson and Company, 1953), p. 227.

17. A.A. Powell and F.H.G. Gruen, "The Constant Elasticity of Transformation Frontier and Linear Supply System," *International Economic Review* 9 (October 1968): 315-328.

18. K.J. Arrow, "The Economic Implications of Learning by Doing," *Review of Economic Studies* 29, 80 (June 1962): 155-73.

19. Derived in K.J. Arrow, H.B. Chenery, B.S. Minhas, and R.M. Solow, "Capital Labor Substitution and Economic Efficiency," *Review of Economics and Statistics* 43 (1961): 225-50.

20. Giora Hanoch, "Production and Demand Models with Direct or Indirect Implicit Additivity," Harvard Institute of Economic Research, Discussion Paper Number 331, November, 1973, p. II-6.

21. Clifford Hildreth and F.J. Jarret, *A Statistical Study of Livestock Production and Marketing* (New York: John Wiley and Sons, 1955), p. 108. The above quotation is reproduced in Marc Nerlove, "The Dynamics of Supply," *The Johns Hopkins University Studies in Historical and Political Science* 76, 2 (1958): 37.

22. Stephen L. McDonald, *Federal Tax Treatment of Income From Oil and Gas* (Washington: The Brookings Institution, 1965), p. 17n.

23. William J. Baumol, "Competitive Pricing and the Centralized Market Place," *Eastern Economic Journal* 1, 1 (January 1974): 11.

24. Marc Nerlove, *Estimation and Identification of Cobb-Douglas Production Functions,* (Chicago: Rand McNally and Company, 1965), Chapter 1.

NOTES TO CHAPTER 6
EMPIRICAL RESULTS AND CONCLUSIONS

1. J. Marshak and W.H. Andrews, "Random Simultaneous Equations and the Theory of Production," *Econometrica* 12 (July–October 1944): 143-205.

2. A. Zellner, J. Kmenta, and J. Dreze, "Specification and Estimation of Cobb-Douglas Production Function Models," *Econometrica* 34 (Oct. 1966): 784-95.

3. Irving Hoch, "Simultaneous Equation Bias in the Context of the Cobb-Douglas Production Function," *Econometrica* 26 (Oct. 1958): 566-78.

4. Marc Nerlove, *Estimation and Identification of Cobb-Douglas Production Functions* (Chicago: Rand McNally and Company, 1965), Chapter 1.

5. See Stephen M. Goldfeld, and Richard E. Quandt, *Nonlinear Methods in Econometrics* (Amsterdam: North-Holland Publishing Company, 1971), pp. 233-34.

6. Ibid. A description of these methods may be found in Chapter 1.

7. Phoebus J. Dhrymes, *Econometrics: Statistical Foundations and Applications* (New York: Harper and Row, 1970), pp. 116-30.

8. Ibid., p. 134.

9. For the theoretical development of this procedure, see Marc Nerlove, "The Dynamics of Supply," *Johns Hopkins University Studies in Historical and Political Science* 76, 2 (1958): Chapter 2.

10. See Phoebus J. Dhrymes, *Distributed Lags: Problems of Estimation and Formulation* (San Francisco: Holden-Day, Inc., 1971), p. 101.

11. Ibid, pp. 59–65.

12. This is the procedure suggested by Goldfeld and Quandt, *Nonlinear Methods,* p. 184.

13. The results of that study are reported in Dennis Epple, "Specification and Testing of Error Structures in Nonlinear Simultaneous Equation Models," in Stephen M. Goldfeld and Richard E. Quandt (eds.), *Studies in Nonlinear Estimation* (Cambridge, Mass: Ballinger Publishing Company, forthcoming): Chapter 5.

NOTES TO APPENDIX A

1. A more extensive discussion of these issues is provided in Dennis Epple, "Specification and Testing of Error Structures in Nonlinear Simultaneous Equation Models," in Stephen M. Goldfeld and Richard E. Quandt (eds.) *Studies in Nonlinear Estimation* (Cambridge, Mass.: Ballinger Publishing Company, forthcoming): Chapter 5.

2. The Monte Carlo model and results are discussed in Epple, "Specification and Testing."

NOTES TO APPENDIX B

1. *Joint Association Of Industry Drilling Costs* (American Petroleum Institute, Washington, Annual).

2. Ibid.

Bibliography

BOOKS

Adelman, M.A. *The World Petroleum Market.* Baltimore: The Johns Hopkins University Press, 1972.

Allen, R.G.D. *Mathematical Analysis for Economists.* London: Macmillan and Co., 1964.

Allvine, Fred C. and James M. Patterson. *Competition Ltd.: The Marketing of Gasoline.* Bloomington: Indiana University Press, 1972.

Bradley, Paul G. *The Economics of Crude Petroleum Production.* Amsterdam: North-Holland Publishing Company, 1967.

Bridge, J.L. *Applied Econometrics.* Amsterdam: North-Holland Publishing Company, 1971.

Buckley, Stuart E., et al. *Petroleum Conservation.* Dallas: E.J. Storm Printing Company, 1951.

CONSAD Corporation. *The Economic Factors Affecting the Level of Domestic Petroleum Reserves.* In Part 4 of U.S. Congress. Committee on Ways and Means and Senate Committee on Finance, *Tax Reform Studies and Proposals,* U.S. Treasury Department. Washington: U.S. Government Printing Office, 1969.

deChazeau, Melvin G. and Alfred E. Kahn. *Integration and Competition in the Petroleum Industry.* New Haven: Yale University Press, 1959.

Dhrymes, Phoebus J. *Distributed Lags: Problems of Estimation and Formulation.* San Francisco: Holden-Day, Inc., 1971.

Dhrymes, Phoebus J. *Econometrics: Statistical Foundations and Applications.* New York: Harper and Row, 1970.

Fisher, Franklin M. *Supply and Costs in the U.S. Petroleum Industry: Two Econometric Studies.* Baltimore: The Johns Hopkins Press, 1964.

Gelfand, I.M. and S.V. Fomin. *Calculus of Variations.* Englewood Cliffs, N.J.: Prentice-Hall, Inc., 1963.

Goldfeld, Stephen M. and Richard E. Quandt. *Nonlinear Methods in Econometrics.* Amsterdam: North-Holland Publishing Company, 1971.

Goodier, J. Leslie. *U.S. Federal and Seacoastal State Offshore Mining Laws.* Washington: Nautilus Press, 1972.

Hildreth, Clifford and F.J. Jarret. *A Statistical Study of Livestock Production and Marketing.* New York: John Wiley and Sons, 1955.

Homan, Paul T. and Wallace F. Lovejoy. *Economic Aspects of Oil Conservation Regulation.* Baltimore: The Johns Hopkins Press, 1967.

Homan, Paul T. and Wallace F. Lovejoy. *Methods of Estimating Reserves of Crude Oil, Natural Gas, and Natural Gas Liquids.* Baltimore: The Johns Hopkins Press, 1965.

Interstate Oil Compact Commission. *A Study of Conservation of Oil and Gas in The United States.* Oklahoma City: IOCC, 1964.

Intriligator, Michael D. *Mathematical Optimization and Economic Theory.* Englewood Cliffs, N.J.: Prentice-Hall, Inc., 1971.

Johnston, John. *Econometric Methods,* (2nd ed.). New York: McGraw-Hill, 1971.

Kaufman, Gordon M. *Statistical Decisions and Related Techniques in Oil and Gas Exploration.* Englewood Cliffs: Prentice-Hall, 1963.

Klein, L.R. *A Textbook of Econometrics.* Evanston: Row, Peterson and Company, 1953.

McDonald, Stephen L. *Federal Tax Treatment of Income from Oil and Gas.* Washington: The Brookings Institution, 1965.

McDonald, Stephen L. *Petroleum Conservation in the United States: An Economic Analysis.* Baltimore: The Johns Hopkins Press, 1971.

McLean, John G. and Robert Wm. Haigh. *The Growth of Integrated Oil Companies.* Boston: Division of Research, Harvard University, Graduate School of Business Administration, 1954.

National Petroleum Council. *The Impact of New Technology on the U.S. Petroleum Industry 1946–65.* Washington: The National Petroleum Council, 1967.

Nerlove, Marc. *Estimation and Identification of Cobb-Douglas Production Functions.* Chicago: Rand McNally and Company, 1965.

Public Land Law Review Commission. *One Third the Nations Land.* Washington: U.S. Government Printing Office, 1970.

Report of the Attorney General Pursuant to Section 2 of the Joint Resolution of July 28, 1955 Consenting to an Interstate Compact to Conserve Oil and Gas. Washington: U.S. Government Printing Office, 1957.

Russel, R. Dana. "The Evolution of Exploration Technology." In National Petroleum Council, *Impact of New Technology on the U.S. Petroleum Industry 1946–65.* Washington: National Petroleum Council, 1967.

U.S. Congress, Senate, Committee on Interior and Insular Affairs. *Trends in Oil and Gas Exploration.* Washington: U.S. Government Printing Office, 1972, Volume 1.

Zimmerman, Erich W. *Conservation in the Production of Petroleum.* New Haven: Yale University Press, 1957.

JOURNAL ARTICLES AND MANUSCRIPTS

Arps, J.J. "Analysis of Decline Curves." *Petroleum Development and Technology.* New York: American Institute of Mining and Metallurgical Engineers, 160, 1945.

Arrow, K.J.; H.B. Chenery; B.S. Minhas; and R.M. Solow. "Capital-Labor Substitution and Economic Efficiency." *Review of Economics and Statistics* 43 (1961): 225–50.

Arrow, K.J. "The Economic Implications of Learning by Doing." *Review of Economic Studies* 29, 80 (June 1962): 155–173.

Adelman, Morris A. "Efficiency of Resource Use in Crude Petroleum." *Southern Economic Journal* 31, 2 (October 1964): 107–122.

Baumol, William J. "Competitive Pricing and the Centralized Market Place." *Eastern Economic Journal* 1, 1 (January 1974).

Bradley, Paul G. and R.S. Uhler. "A Stochastic Model for Determining the Economic Prospects of Petroleum Exploration Over Large Regions." *Journal of the American Statistical Association* 65, 330 (June 1970): 623–630.

Christensen, L.R.; D.W. Jorgenson; and L.J. Lau. "Transcendental Logarithmic Production Frontiers." *Review of Economics and Statistics* 4, 1 (February 1973): 28–45.

Cummins, R.G. and R.G. Kuller. "An Economic Model of Production and Investment for Petroleum Reservoirs." *American Economic Review* 64, 1 (March 1974): 66–79.

Dam, Kenneth W. "Implementation of Import Quotas: The Case of Oil." *The Journal of Law and Economics* 14, 1 (April 1971).

Diewert, W.E. "Functional Forms for Profit and Transformation Functions." *Journal of Economic Theory* 6 (1973): 284–316.

Erickson, Edward W. "Economic Incentives, Industrial Structure and the Supply of Crude Oil Discoveries in the U.S., 1946–58/59," (unpublished manuscript).

Erickson, Edward W. and Robert M. Spann. "Supply Response in a Regulated Industry: The Case of Natural Gas." *Bell Journal of Economics and Management Science* 2, 1 (Spring 1971): 94–121.

Gordon, Richard L. "A Reinterpretation of the Pure Theory of Exhaustion." *Journal of Political Economy* 75 (June 1967): 274–286.

Gray, Lewis C. "Rent Under the Assumption of Exhaustibility." *Quarterly Journal of Economics.* Cambridge: Harvard University Press, May 1914.

Griliches, Zvi and Dale W. Jorgensen. "Issues in Growth Accounting: A Reply to Edward F. Denisen." In *Survey of Current Business* 52, 5, Part II (May 1972): 65–94.

Hanoch, Giora. "Production and Demand Models with Direct or Indirect Implicit Additivity." Harvard Institute of Economic Research, Discussion Paper Number 331, November 1973.

Herfindahl, Orris C. "Depletion and Economic Theory." In Mason Gaffney (ed.), *Extractive Resources and Taxation.* Madison: The University of Wisconsin Press, 1967, pp. 63–90.

Hoch, Irving. "Simultaneous Equation Bias in the Context of the Cobb-Douglas Production Function." *Econometrica* 26 (October 1958): 566–578.

Hotelling, Harold. "The Economics of Exhaustible Resources." *Journal of Political Economy* 39, 2 (April 1931): 137–175.

Jorgenson, Dale W. "Capital Theory and Investment Behavior." *American Economic Review* 3 (May 1963): 247–257.

Kahn, Alfred E. "The Oil Depletion Allowance in the Context of Cartelization." *American Economic Review* 54, 4 (June 1964): 286–314.

Khazzoom, J.D. "The FPC Staff's Econometric Model of Natural Gas Supply in the United States." *Bell Journal of Economics and Management Science* 2, 1 (Spring 1971): 51–93.

MacAvoy, Paul W. "The Regulation Induced Shortage of Natural Gas." *Journal of Law and Economics* 14, 1 (April 1971): 167–199.

MacAvoy, Paul W. and Robert S. Pindyck. "Alternative Regulatory Policies for Dealing with the Natural Gas Shortage." *Bell Journal of Economics and Management Science* 4, 2 (Autumn 1973): 454–498.

Mancke, Richard B. "The Longrun Supply Curve of Crude Oil Produced in the United States." *Antitrust Bulletin* 15 (Winter 1970): 727–756.

"Market Demand Proration Has Outlived its Purpose." *Oil and Gas Journal* 70, 11 (March 13, 1972): 19.

Marshak, J. and W.H. Andrews. "Random Simultaneous Equations and the Theory of Production." *Econometrica* 12 (July-Oct. 1944): 143–205.

McDonal, Frank J. "Geophysics." In National Petroleum Council, *Impact of New Technology on the U.S. Petroleum Industry 1946–65*. Washington: National Petroleum Council, 1967.

McDonald, Stephen L. "Percentage Depletion and the Allocation of Resources: The Case of Oil and Gas." *National Tax Journal* 14, 4 (December 1961): 323–336.

McDonald, Stephen L. and James W. McKie. "Petroleum Conservation in Theory and Practice." *Quarterly Journal of Economics* 76 (February 1962): 98–121.

McKie, James W. "Market Structure and Uncertainty in Oil and Gas Exploration." *Quarterly Journal of Economics* 74, 4 (December 1960): 543–571.

Muth, Richard F. "The Derived Demand Curve for a Productive Factor and the Industry Supply Curve." *Oxford Economic Papers*, (New Series), 16, 2 (July 1964): 221–234.

Nerlove, Marc. "The Dynamics of Supply." *The Johns Hopkins University Studies in Historical and Political Science* 76, 2 (1958).

Nerlove, Marc. "Recent Empirical Studies of the CES and Related Production Functions." In Murray Brown (ed.), *The Theory and Empirical Analysis of Production*. New York: National Bureau of Economic Research, 1967.

Peterson, Frederick M. *The Theory of Exhaustible Natural Resources: A Classical Variational Approach*. Ph.D. dissertation, Princeton University, 1972.

Pindyck, Robert S. "The Regulatory Implications of Three Alternative Econometric Models of Natural Gas Supply." *Bell Journal of Economics and Management Science* 5, 2 (Autumn 1974): 633–645.

Powell, A.A. and F.H.G. Gruen. "The Constant Elasticity of Transformation Frontier and Linear Supply System." *International Economic Review* 9 (October 1968): 315–328.

Reinhardt, Uwe. *An Economic Analysis of Physicians Practices*. Ph.D. dissertation, Yale University, 1970.

Scott, Anthony. "The Theory of the Mine Under Conditions of Certainty." In Mason Gaffney (ed.), *Extractive Resources and Taxation.* Madison: The University of Wisconsin Press, 1967, pp. 25-62.

Steele, H.B. U.S. Congress, Senate, Committee on the Judiciary. Subcommittee on Antitrust and Monopoly. *Government Intervention in the Market Mechanism: The Petroleum Industry, Part 1, Economists Views.* Washington: U.S. Government Printing Office, 1969, pp. 208-233.

VanDyke, L.H. "North American Drilling Activity in 1967." *Bulletin of the American Association of Petroleum Geologists* 52, 6 (June 1968): 895-926.

Walters, A.A. "Production and Cost Functions: An Econometric Survey." *Econometrica* 31, 1-2 (Jan.-Apr. 1963): 1-66.

Zellner, A.; J. Kmenta; and J. Dreze. "Specification and Estimation of Cobb-Douglas Production Function Models." *Econometrica* 34 (Oct. 1966): 784-795.

DATA SOURCES

Annual Statistical Review: U.S. Petroleum Industry Statistics 1956-1972. Washington: American Petroleum Institute, April 1973.

Bulletin of The American Association of Petroleum Geologists, June issues.

Independent Petroleum Association of America. *The Oil Producing Industry in Your State,* 1972.

Independent Petroleum Association of America. *The Petroleum Independent,* Sept./Oct. 1971.

Joint Association of the U.S. Oil and Gas Producing Industry sponsored by the American Petroleum Institute, the Independent Petroleum Association of America, and the Mid-continent Oil and Gas Association. Washington: The American Petroleum Institute, 1972.

Petroleum Facts and Figures. Washington: American Petroleum Institute, 1971.

Report of The Cost Study Committee, Washington: The Independent Petroleum Association of America, various years.

Reserves of Crude Oil, Natural Gas Liquids, And Natural Gas in the United States and Canada and United States Producive Capacity as of December 31, 1972. Washington, 1973. Prepared jointly by the American Petroleum Institute, The American Gas Association, and The Canadian Petroleum Association, Washington, 1973.

U.S. Bureau of Mines. *Minerals Yearbook.* Washington: U.S. Government Printing Office, Annual.

World Oil, February 15, 1959, pp. 118-19.

Index

About the Author

Dennis Epple is an Assistant Professor in the Graduate School of Industrial Administration at Carnegie-Mellon University. He received a Ph.D. degree in Economics from Princeton University in January of 1975. The author's research centers on the application of microeconomic techniques and quantitative methods in the analysis of public policy issues with particular emphasis on energy modeling and forecasting.